Symmetries of Nonlinear PDEs on Metric Graphs and Branched Networks

Symmetries of Nonlinear PDEs on Metric Graphs and Branched Networks

Special Issue Editors

Diego Noja

Dmitry E. Pelinovsky

MDPI • Basel • Beijing • Wuhan • Barcelona • Belgrade

Special Issue Editors
Diego Noja
Università di Milano Bicocca
Italy

Dmitry E. Pelinovsky
McMaster University
Canada

Editorial Office
MDPI
St. Alban-Anlage 66
4052 Basel, Switzerland

This is a reprint of articles from the Special Issue published online in the open access journal *Symmetry* (ISSN 2073-8994) in 2019 (available at: https://www.mdpi.com/journal/symmetry/special_issues/Symmetries_Nonlinear_PDEs_Metric_Graphs_Branched_Networks).

For citation purposes, cite each article independently as indicated on the article page online and as indicated below:

LastName, A.A.; LastName, B.B.; LastName, C.C. Article Title. *Journal Name* **Year**, *Article Number*, Page Range.

ISBN 978-3-03921-720-5 (Pbk)
ISBN 978-3-03921-721-2 (PDF)

Contents

About the Special Issue Editors

Diego Noja is Professor of Mathematical Physics at University of Milano Bicocca (Italy). He completed his Ph.D. in 1997 in Milano. His research is related to two main issues. The first is the behavior of classical and quantum fields interacting with particles, in particular, as regards the well-posedness of dynamics as well as infrared and ultraviolet problems. The second is the analysis of dispersive Hamiltonian equations in the presence of point defects and on metric graphs.

Dmitry E. Pelinovsky is Professor of Mathematics at McMaster University (Canada). He was born in Russia and completed his Ph.D. in 1997 at Monash University (Australia). His research is centered on the problems of stability of nonlinear waves. He is an author of the monograph "Localization in Periodic Potentials: From Schrodinger Operators to the Gross–Pitaevskii equation" (Cambridge University Press, 2011), Editor of "Nonlinear Physical Systems: Spectral Analysis, Stability and Bifurcations" (Wiley, 2014), and Guest Editor of Special Issues of *Chaos* (2005), *Applicable Analysis* (2010), *Discrete Continuous Dynamical Systems* (2012), and *Journal of Dynamics and Differential Equations* (2020). He has written more than 250 papers published in peer-reviewed journals. He is Editor of *Physica D* and *Studies in Applied Mathematics*.

Preface to "Symmetries of Nonlinear PDEs on Metric Graphs and Branched Networks"

This Special Issue focuses on recent progress in a new area of mathematical physics and applied analysis, namely, on nonlinear partial differential equations on metric graphs and branched networks. Graphs represent a system of edges connected at one or more branching points (vertices). The connection rule determines the graph topology. When the edges can be assigned a length and the wave functions on the edges are defined in metric spaces, the graph is called a metric graph.

Evolution equations on metric graphs have attracted much attention as effective tools for the modeling of particle and wave dynamics in branched structures and networks. Since branched structures and networks appear in different areas of contemporary physics with many applications in electronics, biology, material science, and nanotechnology, the development of effective modeling tools is important for the many practical problems arising in these areas.

The list of important problems includes searches for standing waves, exploring of their properties (e.g., stability and asymptotic behavior), and scattering dynamics. This Special Issue is a representative sample of the works devoted to the solutions of these and other problems. The contributions to the Special Issue are written by selected participants of the workshop "Nonlinear PDEs on Metric Graphs and Branched Networks" organized at Lorentz center, Leiden, the Netherlands, in August 2018.

Diego Noja, Dmitry E. Pelinovsky
Special Issue Editors

symmetry

MDPI

Article

Approximations of Metric Graphs by Thick Graphs and Their Laplacians

Olaf Post

Fachbereich 4—Mathematik, Universität Trier, 54286 Trier, Germany; olaf.post@uni-trier.de

Received: 30 January 2019; Accepted: 8 March 2019; Published: 12 March 2019

Abstract: The main purpose of this article is two-fold: first, to justify the choice of Kirchhoff vertex conditions on a metric graph as they appear naturally as a limit of Neumann Laplacians on a family of open sets shrinking to the metric graph ("thick graphs") in a self-contained presentation; second, to show that the metric graph example is close to a physically more realistic model where the edges have a thin, but positive thickness. The tool used is a generalization of norm resolvent convergence to the case when the underlying spaces vary. Finally, we give some hints about how to extend these convergence results to some mild non-linear operators.

Keywords: metric graphs; open sets converging to metric graphs; Laplacians; norm convergence of operators; convergence of spectra

1. Introduction

The study of operators on metric graphs has been an ongoing and active area of research for at least two decades. Several natural questions arise in the study of Laplacians on metric graphs: As there is some freedom in defining a *self-adjoint* Laplacian on a metric graph due to the vertex conditions (see, e.g., [1] and the references therein), can one justify a certain choice of such vertex conditions? Second, in a realistic physical model (a *thick graph*), the wires have a thickness of order ε, but in the metric graph model, it is simplified to $\varepsilon = 0$: Can one justify some sort of limit of a Laplacian on the network with thickness $\varepsilon > 0$ as $\varepsilon \to 0$?

The aim of this article is to give an answer to both questions. We show that the Neumann Laplacian on the ε-neighborhood of the metric graph (embedded in some ambient space $\mathbb{R}^{m+1}$) converges to the Kirchhoff Laplacian on the metric graph. This gives answers to both questions above: First, the "natural" vertex conditions are the so-called *Kirchhoff* conditions; see Equations (3) and (4). Second, the limit problem is a good approximation to a realistic physical model on a thick graph as $\varepsilon \to 0$. Note that the problem significantly simplifies in the limit, as we only have to consider a system of ODEs instead of a PDE on a complicated and ε-dependent space. Moreover, the problem on the metric graph can often be solved explicitly.

A technical difficulty is that the Laplacian on the thick graph and on the metric graph live on different spaces. We therefore generalize the notion of norm resolvent convergence to this case; this was first done in [2]; see also the monograph [3] for a history of the problem and [4] for a recent list of references. Convergence of the (discrete) spectrum for the Neumann Laplacian on a thick graph converging to a compact metric graph has already been established by variational methods in [5–7].

The aim of this article is also to provide an almost self-contained presentation of the results for linear operators on thick and metric graphs to the "non-linear" community and also to give some ideas of how they can be extended to some mild non-linear operators.

2. Metric Graphs and Their Laplacians

For a detailed presentation of metric graphs and their Laplacians, we refer to [1,3] and the references therein. Let X_0 denote a metric graph given by the data (V, E, ℓ), where V and E are the (at most countable) sets of *vertices* and *edges*, respectively, and where $\ell \colon E \longrightarrow (0, \infty)$. $e \mapsto \ell_e$ denotes the *length* of the edge $e \in E$; a *metric edge* will be the interval $I_e := [0, \ell_e]$. The *metric graph* X_0 is now the disjoint union of all metric edges $\dot{\bigcup}_{e \in E} I_e$ after identifying the endpoints ∂I_e with the corresponding vertices. A metric graph is a metric space using the intrinsic metric (i.e., $d(s, \widetilde{s})$ is the length of the shortest path in X_0 between s and $\widetilde{s}$). Moreover, there is a natural measure on X_0 given by the sum of the Lebesgue measures on each metric edge I_e.

As the *Hilbert space* on X_0, we choose:

$$\mathscr{H}_0 := \mathsf{L}_2(X_0) = \bigoplus_{e \in E} \mathsf{L}_2(I_e),$$

where we write $f \in \mathsf{L}_2(X_0)$ as family $(f_e)_{e \in E}$ with $f_e \in \mathsf{L}_2(I_e)$; moreover, $\bigoplus_{e \in E} \mathsf{L}_2(I_e)$ denotes the Hilbert orthogonal sum with f being in it if its squared norm:

$$\|f\|^2_{\mathsf{L}_2(X_0)} := \sum_{e \in E} \int_{I_e} |f_e(s)|^2 \, \mathrm{d}s$$

is finite. Similarly, we define $\mathsf{H}^k_{\mathrm{dec}}(X) := \bigoplus_{e \in E} \mathsf{H}^k(I_e)$ for $k \in \mathbb{N}_0$. The label "dec" refers to the fact that for $k \geq 1$, there is no relation between the (well-defined) values of f_e and its derivatives at a vertex v for different $e \in E_v$. Here, E_v denotes the set of edges that are adjacent with the vertex $v \in V$. Recall that functions in $\mathsf{H}^1(I_e)$ are continuous as we have the estimate:

$$\left|f_e(s) - f_e(\widetilde{s})\right|^2 \leq |s - \widetilde{s}| \int_{I_e} |f_e'(u)|^2 \, \mathrm{d}u.$$

Using a suitable cut-off function, we conclude the Sobolev trace estimate:

$$|f_e(v)|^2 \leq C_e \|f_e\|_{\mathsf{H}^1(I_e)} = C_e \int_{I_e} \left(|f_e(s)|^2 + |f_e'(s)|^2\right) \mathrm{d}s \tag{1}$$

with $C_e = 2/\min\{1, \ell_e\}$, where $f_e(v)$ denotes the evaluation of f_e at one of the endpoints of I_e corresponding to $v \in V$. In particular, we assume that:

$$\ell_0 := \min\{\inf_{e \in E} \ell_e, 1\} > 0. \tag{2}$$

From (1) and (2), we then conclude that the subspace:

$$\mathsf{H}^1(X_0) := \mathsf{H}^1_{\mathrm{dec}}(X_0) \cap \mathsf{C}(X_0) = \left\{ f \in \mathsf{H}^1_{\mathrm{dec}}(X_0) \mid f_e(v) \text{ is independent of } e \in E_v \text{ for all } v \in V \right\} \tag{3}$$

is closed in $\mathsf{H}^1_{\mathrm{dec}}(X_0)$. We denote by $f(v) := f_e(v)$ the common value of f at the vertex v. It follows that:

$$\mathfrak{l}_0(f) := \|f'\|^2 = \sum_{e \in E} \int_{I_e} |f_e'(s)|^2 \, \mathrm{d}s, \qquad f \in \operatorname{dom} \mathfrak{l}_0 := \mathsf{H}^1(X_0),$$

defines a closed, non-negative quadratic form in $\mathscr{H}_0 = \mathsf{L}_2(X_0)$. The associated self-adjoint and non-negative operator L_0 is given by:

$$(L_0 f)_e = -f_e'', \qquad f \in \operatorname{dom} L_0 = \left\{ f \in \mathsf{H}^2_{\mathrm{dec}}(X_0) \,\middle|\, f \in \mathsf{C}(X_0), \sum_{e \in E_v} f_e'(v) = 0 \right\}. \tag{4}$$

Here, $f_e'(v)$ denotes the (weak) derivative of f_e along e *towards* the vertex v. The operator L_0 is sometimes referred to as the *(generalized) Neumann Laplacian* or *Kirchhoff Laplacian* (the second because of the flux condition $\sum_{e\in E_v} f_e'(v) = 0$ on the *derivatives*). Note that for vertices of degree one, the vertex condition is just the usual Neumann boundary condition $f_e'(v) = 0$, and for vertices of degree two, we have $f_{e_1}(v) = f_{e_2}(v)$ and $f_{e_1}'(v) + f_{e_2}'(v) = 0$, i.e., the continuity of f and its derivative along v (recall that $f_e'(v)$ denotes the derivative *towards* the vertex v).

3. Thick Graphs and Their Laplacians

We assume first that the metric graph X_0 is embedded in some space $\mathbb{R}^{m+1}$ ($m \geq 1$) such that all edges are straight line segments in $\mathbb{R}^{m+1}$. For $\varepsilon > 0$, denote by:

$$X_\varepsilon^\diamond := \{\, x \in \mathbb{R}^{m+1} \,|\, d(x, X_0) < \varepsilon/\omega_m \,\}$$

the ε/ω_m-neighborhood of X_0 in $\mathbb{R}^{m+1}$. Here, ω_m is the mth root of the volume of the unit Euclidean ball in $\mathbb{R}^m$, i.e., $\omega_1 = 2$, $\omega_2 = \sqrt{\pi}$, $\omega_3 = \sqrt[3]{4\pi/3}$, etc. We say that $X_\varepsilon^\diamond$ is a *graph-like space* or a *thick graph* constructed from the metric graph X_0 if there is $\varepsilon_0 > 0$ such that:

$$\overline{X_\varepsilon^\diamond} = \bigcup_{v\in V} \overline{X_{\varepsilon,v}^\diamond} \cup \bigcup_{e\in E} \overline{X_{\varepsilon,e}^\diamond} \tag{5}$$

for all $\varepsilon \in (0, \varepsilon_0]$ (cf. Figure 1), where $X_{\varepsilon,v}^\diamond$ and $X_{\varepsilon,e}^\diamond$ are open and pairwise disjoint subsets of $\mathbb{R}^{m+1}$ such that the so-called *vertex* and *edge neighborhoods* fulfil:

$$X_{\varepsilon,v}^\diamond \cong \varepsilon X_v \qquad \text{and} \qquad X_{\varepsilon,e}^\diamond \cong (\varepsilon, \ell_e - 2a_e\varepsilon) \times \varepsilon B, \tag{6}$$

i.e., $X_{\varepsilon,v}^\diamond$ is isometric to the ε-scaled version of an open subset X_v, $X_{\varepsilon,e}^\diamond$ is isometric with the product of an interval of length $\ell_e - 2a_e\varepsilon$, and $B \subset \mathbb{R}^m$ is a ball of radius $1/\omega_m$, having m-dimensional volume one by the definition of the scaling factor ω_m. Moreover, $2a_e\varepsilon$ is the sum of the lengths of the two parts of the metric edge inside the vertex neighborhood. For finite graphs, the existence of $\varepsilon_0 > 0$ is no restriction, but for infinite graphs with an arbitrary large vertex degree, this might be a restriction on the embedding and the edge lengths. More details on spaces constructed according to a graph (so-called "graph-like spaces") can be found in the monograph [3]; see also the references therein.

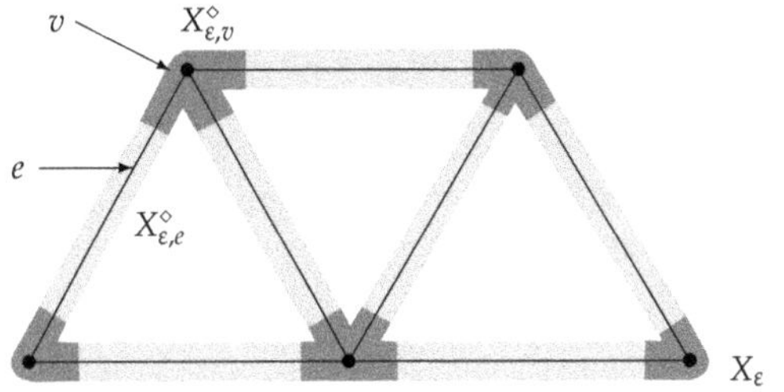

Figure 1. The decomposition of a graph-like space of thickness of order ε into vertex neighborhoods $X_{\varepsilon,v}^\diamond$ (dark grey) and edge neighborhoods $X_{\varepsilon,e}^\diamond$ (light grey) according to a metric graph X_0 embedded in $\mathbb{R}^2$.

As the Hilbert space, we set $\mathscr{H}_\varepsilon^\diamond := \mathsf{L}_2(X_\varepsilon^\diamond)$. As the operator, we use the (non-negative) Neumann Laplacian $L_\varepsilon^\diamond$ defined as the self-adjoint and non-negative operator associated with the closed and non-negative quadratic form given by:

$$\mathfrak{a}_\varepsilon^\diamond(u) := \|\nabla u\|^2_{\mathsf{L}_2(X_\varepsilon^\diamond)}$$

in $\mathscr{H}_\varepsilon^\diamond = \mathsf{L}_2(X_\varepsilon^\diamond)$.

In our calculations later, it is more convenient to work with edge neighborhoods $X_{\varepsilon,e}$ that are isometric with the product of the *original* edge I_e times the ε-scaled ball B, i.e.,

$$X_{\varepsilon,e} \cong I_e \times \varepsilon B$$

instead of the slightly shortened edge neighborhood $X_{\varepsilon,e}^\diamond$. For $v \in V$, we set $X_{\varepsilon,v} = X_{\varepsilon,v}^\diamond = \varepsilon X_v$. We then construct X_ε as the space obtained from gluing the building blocks $X_{\varepsilon,v}$ and $X_{\varepsilon,e}$ such that a decomposition similar to (5) holds, now without the label $(\cdot)^\diamond$. Note that X_ε is defined as an abstract flat manifold with boundary and might not be embeddable into $\mathbb{R}^{m+1}$ any longer. We also call X_ε a *graph-like space* or *thick graph*. We state that the Neumann Laplacians on X_ε and $X_\varepsilon^\diamond$ are "close to each other" in Lemma 4.

Due to a decomposition of X_ε into its building blocks similar to (5) and the scaling behavior, the norm in the Hilbert space $\mathscr{H}_\varepsilon := \mathsf{L}_2(X_\varepsilon)$ fulfills:

$$\|u\|^2_{\mathsf{L}_2(X_\varepsilon)} = \sum_{v\in V} \varepsilon^{m+1} \int_{X_v} |u_v(x)|^2 \,\mathrm{d}x + \sum_{e\in E} \varepsilon^m \int_{I_e}\int_B |u_e(s,y)|^2 \,\mathrm{d}y\,\mathrm{d}s,$$

where u_v and u_e denote the restriction of u onto the ε-independent building blocks X_v and $X_e = I_e \times B$. Note that with this notation, we have put all ε-dependencies into the norm (and later also into the quadratic form).

As the operator, we use the (negative) Neumann Laplacian L_ε defined as the self-adjoint and non-negative operator associated with the closed and non-negative quadratic form given by:

$$\begin{aligned}\mathfrak{l}_\varepsilon(u) &:= \|\nabla u\|^2_{\mathsf{L}_2(X_\varepsilon)} \\ &= \sum_{v\in V} \varepsilon^{m-1} \int_{X_v} |\nabla u_v(x)|^2 \,\mathrm{d}x + \sum_{e\in E} \varepsilon^m \int_{I_e}\int_B \Big(|u_e'(s,y)|^2 + \frac{1}{\varepsilon^2} |\nabla_B u_e(s,y)|^2 \Big) \,\mathrm{d}y\,\mathrm{d}s\end{aligned}$$

in $\mathscr{H}_\varepsilon = \mathsf{L}_2(X_\varepsilon)$ using the scaling behavior of the building blocks. Here, u_e' denotes the derivative with respect to the longitudinal (first) variable s, and ∇_B denotes the derivative with respect to the second variable $y \in B$.

4. Convergence of the Resolvents

How can we now compare the two Laplacians L_0 and L_ε (resp. $L_\varepsilon^\diamond$)? The idea is first to consider the resolvents:

$$R_0 := (L_0 + 1)^{-1} \qquad \text{resp.} \qquad R_\varepsilon := (L_\varepsilon + 1)^{-1}$$

in $\mathscr{H}_0$, resp. $\mathscr{H}_\varepsilon$, since they are bounded operators. In order to define a norm difference of these resolvents, we need a so-called *identification operator:*

$$J_\varepsilon \colon \mathscr{H}_0 \longrightarrow \mathscr{H}_\varepsilon,$$

in our situation given by

$$(J_\varepsilon f)_v = 0 \qquad \text{and} \qquad (J_\varepsilon f)_e(s,y) = \varepsilon^{-m/2} f_e(s),$$

i.e., we set Jf to zero on the vertex neighborhood and transversally constant on the edge neighborhood, together with an appropriate rescaling constant. As the identification operator in the opposite direction, we use $J_\varepsilon^* \colon \mathscr{H}_\varepsilon \longrightarrow \mathscr{H}_0$, where an easy calculation shows that:

$$(J_\varepsilon^* u)_e(s) = \varepsilon^{m/2} \int_B u_e(s,y)\,\mathrm{d}y.$$

It is easy to see that $J_\varepsilon^* J_\varepsilon f = f$, i.e., J_ε is an isometry.
We now compare the two resolvents, sandwiched with J_ε. Let:

$$D_\varepsilon := R_\varepsilon J_\varepsilon - J_\varepsilon R_0 \colon \mathscr{H}_0 \longrightarrow \mathscr{H}_\varepsilon.$$

What does D_ε look like? The best way to deal with it is to consider $\langle D_\varepsilon g, w\rangle_{\mathscr{H}_\varepsilon}$ for $g \in \mathscr{H}_0$ and $w \in \mathscr{H}_\varepsilon$. We have:

$$\begin{aligned}\langle D_\varepsilon g, w\rangle_{\mathscr{H}_\varepsilon} &= \langle J_\varepsilon g, R_\varepsilon w\rangle_{\mathscr{H}_\varepsilon} - \langle J_\varepsilon R_0 g, w\rangle_{\mathscr{H}_\varepsilon}\\ &= \langle J_\varepsilon L_0 f, u\rangle_{\mathscr{H}_\varepsilon} - \langle J_\varepsilon f, L_\varepsilon u\rangle_{\mathscr{H}_\varepsilon},\end{aligned}$$

where $u = R_\varepsilon w \in \operatorname{dom} L_\varepsilon$ and $f = R_0 g \in \operatorname{dom} L_0$. Note that

$$(L_\varepsilon u)_e = -u_e'' + \frac{1}{\varepsilon^2} L_B u_e,$$

where L_B is (minus) the Neumann Laplacian on B acting on the second variable $y \in B$. In particular, we conclude:

$$\begin{aligned}\langle J_\varepsilon L_0 f, u\rangle_{\mathscr{H}_\varepsilon} - \langle J_\varepsilon f, L_\varepsilon u\rangle_{\mathscr{H}_\varepsilon} &= \varepsilon^{m/2} \sum_{e\in E} \int_{I_e} \int_B \Big(-f_e''(s)\overline{u}_e(s,y) - f_e(s)\Big(-\overline{u}_e'' + \frac{1}{\varepsilon^2} L_B \overline{u}_e\Big)(s,y)\Big)\,\mathrm{d}y\,\mathrm{d}s\\ &= \varepsilon^{m/2} \sum_{e\in E} \Big[-f_e'(s)\int_B \overline{u}_e(s,y)\,\mathrm{d}y + f_e(s)\int_B \overline{u}_e'(s,y)\,\mathrm{d}y\Big]_{s=0}^{\ell_e}\\ &= \varepsilon^{m/2} \sum_{v\in V}\sum_{e\in E_v} \Big(-f_e'(v)\int_B \overline{u}_e(v,y)\,\mathrm{d}y + f_e(v)\int_B \overline{u}_e'(v,y)\,\mathrm{d}y\Big),\end{aligned}$$

where we used partial integration and the fact that L_B is a self-adjoint operator in $\mathsf{L}_2(B)$ and $L_B f_e = 0$ (as f_e is independent of the second variable y) for the second equality and a reordering argument in the third equality. Moreover, plugging v into s means evaluation at $s = 0$, resp. $s = \ell_e$, if v corresponds to zero, resp. ℓ_e; for the longitudinal derivative, we assume $u_e'(v,y) = -u_e'(0,y)$, resp. $u_e'(v,y) = u_e'(\ell_e,y)$ if v corresponds to zero, resp. ℓ_e.

We now use the fact that $f \in \operatorname{dom} L_0$: first note that $\sum_{e\in E_v} f_e'(v) = 0$, so that we can smuggle in a constant $C_v u$ into the first summand, namely:

$$\sum_{e\in E_v}\Big(-f_e'(v)\int_B u_e(v,y)\,\mathrm{d}y\Big) = \sum_{e\in E_v} f_e'(v)\Big(C_v u - \int_B u_e(v,y)\Big)\,\mathrm{d}y.$$

We specify $C_v u$ in a moment. For the second summand, we use the fact that $f_e(v) = f(v)$ is independent of $e \in E_v$, and we have:

$$\sum_{e\in E_v} f_e(v)\int_B u_e'(v,y)\,\mathrm{d}y = -f(v)\int_{\partial X_v} \partial_n u(x)\,\mathrm{d}x = -f(v)\int_{X_v} \Delta u(x)\,\mathrm{d}x.$$

For the second equality, we used the fact that B at $s = v$ corresponds to the subset $\partial_e X_v$ of ∂X_v where the edge neighborhood is attached and that the normal derivative (pointing outwards) of u vanishes on $\partial X_{\varepsilon,v} \cap \partial X_\varepsilon$ due to the Neumann conditions. For the last equality, we used the Gauss–Green formula (write $\partial_n u = \partial_n u \cdot 1$).

As $u \in \operatorname{dom} L_\varepsilon$, we expect that the average $\int_B u(v,y)\,\mathrm{d}y$ of u over the boundary component $\partial_e X_v$ is close to the average of u over X_v itself (recall that $\operatorname{vol}_m B = 1$); hence, we set:

$$C_v u := \fint_{X_v} u(x)\,\mathrm{d}x.$$

Define now:

$$(A_0 g)(v) := (f'_e(v))_{e\in E_v} \in \mathbb{C}^{E_v}, \qquad (A_\varepsilon w)(v) := \varepsilon^{m/2}\Big(C_v u - \int_B u_e(v,y)\,\mathrm{d}y\Big)_{e\in E_v} \in \mathbb{C}^{E_v},$$

$$(B_0 g)(v) := f(v) \in \mathbb{C}, \qquad (B_\varepsilon w)(v) := \frac{\varepsilon^{m/2}}{\deg v}\int_{X_v}(-\Delta u)(x)\,\mathrm{d}x,$$

where $\deg v$ denotes the degree of v (i.e., the number of elements in E_v), then we have shown that:

$$\langle D_\varepsilon g, w\rangle_{\mathscr{H}_\varepsilon} = \sum_{v\in V}\Big(\langle (A_0 g)(v),(A_\varepsilon w)(v)\rangle_{\mathbb{C}^{E_v}} + (B_0 g)(v)\overline{(B_\varepsilon \overline{w})(v)}\deg v\Big).$$

Defining $\mathscr{G} := \ell_2(V,\deg)$ (with the weighted norm given by $\|\varphi\|^2_{\ell_2(V,\deg)} = \sum_{v\in V}|\varphi(v)|^2\deg v$) and $\widetilde{\mathscr{G}} := \bigoplus_{v\in V}\mathbb{C}^{E_v}$, the previous equation reads as:

$$D_\varepsilon = A_\varepsilon^* A_0 + B_\varepsilon^* B_0 \tag{7}$$

in operator notation, where:

$$A_0\colon \mathscr{H}_0 \longrightarrow \widetilde{\mathscr{G}}, \quad (A_0 g)_e(v) = (R_0 g)'_e(v), \qquad B_0\colon \mathscr{H}_0 \longrightarrow \mathscr{G}, \quad (B_0 g)(v) = (R_0 g)(v)$$

and:

$$A_\varepsilon\colon \mathscr{H}_\varepsilon \longrightarrow \widetilde{\mathscr{G}}, \qquad (A_\varepsilon w)_e(v) = \varepsilon^{m/2}\Big(C_v(R_\varepsilon w) - \int_B (R_\varepsilon w)_e(v,y)\,\mathrm{d}y\Big),$$

$$B_\varepsilon\colon \mathscr{H}_\varepsilon \longrightarrow \mathscr{G}, \qquad (B_\varepsilon w)(v) = \frac{\varepsilon^{m/2}}{\deg v}\int_{X_v}(-\Delta(R_\varepsilon w))(x)\,\mathrm{d}x.$$

Let us now estimate the norms of the auxiliary operators: it also explains why we work with the weighted space $\ell_2(V,\deg)$:

Lemma 1. *Assume that (2) holds, then:*

$$\|A_0\|_{\mathscr{H}_0\to\widetilde{\mathscr{G}}} \le \frac{2\sqrt{2}}{\sqrt{\ell_0}} \quad \text{and} \quad \|B_0\|_{\mathscr{H}_0\to\mathscr{G}} \le \frac{2}{\sqrt{\ell_0}}.$$

Proof. From (1) and (2), for each f_e, the fact that $f(v) = f_e(v)$, and summing over $v \in V$, we conclude:

$$\|B_0 g\|^2_{\ell_2(V,\deg)} = \sum_{v\in V}|f(v)|^2\deg v = \sum_{v\in V}\sum_{e\in E_v}|f_e(v)|^2 \le \sum_{v\in V}\sum_{e\in E_v}\frac{2}{\ell_0}\|f_e\|^2_{\mathsf{H}^1(I_e)} = \frac{4}{\ell_0}\sum_{e\in E}\|f_e\|^2_{\mathsf{H}^1(I_e)}$$

where $g = R_0 f$. Now, the last sum equals:

$$\mathfrak{l}_0(f) + \|f\|^2_{\mathscr{H}_0} = \|(L_0+1)^{1/2}f\|^2_{\mathscr{H}_0} = \|(L_0+1)^{-1/2}g\|^2_{\mathscr{H}_0} \le \|g\|^2_{\mathscr{H}_0};$$

hence, the second norm estimate holds. For the first one, we argue: similarly

$$\|A_0 g\|^2_{\widetilde{\mathscr{G}}} = \sum_{v\in V}\sum_{e\in E_v}|f'_e(v)|^2 \le \sum_{v\in V}\sum_{e\in E_v}\frac{2}{\ell_0}\|f'_e\|^2_{\mathsf{H}^1(I_e)} = \frac{4}{\ell_0}\big(\|f'\|^2_{\mathsf{L}_2(X_0)} + \|f''\|^2_{\mathsf{L}_2(X_0)}\big),$$

Now,

$$\|f'\|_{\mathsf{L}_2(X_0)} = \|L_0^{1/2}(L_0+1)^{-1}g\|_{\mathsf{L}_2(X_0)} \le \|g\|_{\mathsf{L}_2(X_0)} \quad \text{and}$$
$$\|f''\|_{\mathsf{L}_2(X_0)} = \|L_0(L_0+1)^{-1}g\|_{\mathsf{L}_2(X_0)} \le \|g\|_{\mathsf{L}_2(X_0)}$$

by the spectral calculus, and the first norm estimate follows. □

More importantly, we now show that the ε-dependent operators have actually a norm converging to zero as $\varepsilon \to 0$:

Lemma 2. *Assume that* (2) *and:*

$$\deg v \le d_0 < \infty, \qquad \lambda_2(X_v) \ge \lambda_2 > 0 \quad \text{and} \quad \frac{\operatorname{vol} X_v}{\deg v} \le c < \infty \tag{8}$$

hold (By some modifications in the decomposition (6) *(namely, one uses $X^{\diamond}_{\varepsilon,e} = (\varepsilon a_e, \ell_e - a_e\varepsilon)$ for some appropriate $a_e > 0$), one can avoid a direct upper bound d_0 on the vertex degrees, but then a_e has to be large if $\deg v$ is large; also, the high degree will make $\operatorname{vol} X_v$ larger in order to have enough space to attach all the edge neighborhood; see also the discussion in ([2], Section 3.1.) for all $v \in V$, where $\lambda_2(X_v)$ is the second (first non-zero) Neumann eigenvalue of X_v, then:*

$$\|A_\varepsilon\|^2_{\mathscr{H}_\varepsilon \to \widetilde{\mathscr{G}}} \le 2\varepsilon d_0\Big(1 + \frac{1}{\lambda_2}\Big) \qquad \text{and} \qquad \|B_\varepsilon\|^2_{\mathscr{H}_\varepsilon \to \mathscr{G}} \le \varepsilon^3 c.$$

Proof. We need the following vector-valued version of (1):

$$\|u_e(v,\cdot)\|^2_{\mathsf{L}_2(B)} \le 2(\|\nabla u\|^2_{\mathsf{L}_2(X_v)} + \|u\|^2_{\mathsf{L}_2(X_v)}) \tag{9}$$

(actually, we apply (1) to $u(\cdot,y)$ for each $y \in B$ into a line of length one at $y \in B$ perpendicular to $\partial_e X_v \cong B$ into X_v, and integrate then over $y \in B$). We then have:

$$\begin{aligned}\Big|C_v u - \int_B u_e(v,y)\,\mathrm{d}y\Big|^2 = \Big|\int_B (C_v u - u_e(v,y))\,\mathrm{d}y\Big|^2 &\le \|C_v u - u_e\|^2_{\mathsf{L}_2(B)} \\ &\le 2(\|\nabla u\|^2_{\mathsf{L}_2(X_v)} + \|u - C_v u\|^2_{\mathsf{L}_2(X_v)})\end{aligned}$$

(recall that $\int_B \mathrm{d}x = 1$). Now, $u - C_v u$ is the projection onto the eigenspace of the Neumann problem on X_v of all eigenfunctions orthogonal to the constant; hence, we have:

$$\|u - C_v u\|^2_{\mathsf{L}_2(X_v)} \le \frac{1}{\lambda_2(X_v)}\|\nabla u\|^2_{\mathsf{L}_2(X_v)}$$

by the variational characterization of eigenvalues. In particular, we have:

$$\Big|C_v u - \int_B u_e(v,y)\,\mathrm{d}y\Big|^2 \le 2\Big(1 + \frac{1}{\lambda_v(X_v)}\Big)\|\nabla u\|^2_{\mathsf{L}_2(X_v)}.$$

Now, letting $u = R_\varepsilon w$, we have:

$$\|A_\varepsilon w\|^2_{\widetilde{\mathscr{G}}} = \varepsilon^m \sum_{v\in V}\sum_{e\in E_v}\Big|C_v u - \int_B u_e(v,y)\,\mathrm{d}y\Big|^2 \le 2\varepsilon^m \sum_{v\in V} \deg v\Big(1 + \frac{1}{\lambda_v(X_v)}\Big)\|\nabla u\|^2_{\mathsf{L}_2(X_v)}.$$

Moreover,

$$\varepsilon^m \sum_{v\in V}\|\nabla u\|^2_{\mathsf{L}_2(X_v)} \le \varepsilon\mathfrak{l}_\varepsilon(u) = \varepsilon\|L_\varepsilon^{1/2}(L_\varepsilon+1)^{-1}w\|^2_{\mathsf{L}_2(X_\varepsilon)} \le \varepsilon\|w\|^2_{\mathsf{L}_2(X_\varepsilon)};$$

hence, $\|A_\varepsilon\|^2_{\mathscr{H}_\varepsilon \to \widetilde{\mathscr{G}}} \le \varepsilon \sup_{v\in V} 2(\deg v)(1 + 1/\lambda_2(X_v)) \le 2\varepsilon d_0(1 + 1/\lambda_2)$.

For the second norm estimate, we have:

$$\begin{aligned}\|B_\varepsilon w\|^2_{\mathscr{G}} = \varepsilon^m \sum_{v\in V} \frac{1}{\deg v}\left|\int_{X_v}(-\Delta u)(x)\,\mathrm{d}x\right|^2 &\le \varepsilon^m \sum_{v\in V}\frac{\operatorname{vol} X_v}{\deg v}\|-\Delta w\|^2_{\mathsf{L}_2(X_v)}\\ &\le \varepsilon^3 c\|L_\varepsilon(L_\varepsilon+1)^{-1}w\|^2_{\mathsf{L}_2(X_\varepsilon)} \le \varepsilon^3 c\|w\|^2_{\mathsf{L}_2(X_\varepsilon)} \quad \square\end{aligned}$$

From the calculation of D_ε in (7) and Lemmas 1 and 2, we conclude:

Theorem 1. *Under the uniformity assumptions* (2) *and* (8), *the operator norm of:*

$$D_\varepsilon = R_\varepsilon J_\varepsilon - J_\varepsilon R_0 \colon \mathscr{H}_0 = \mathsf{L}_2(X_0) \longrightarrow \mathscr{H}_\varepsilon = \mathsf{L}_2(X_\varepsilon)$$

is of order $\varepsilon^{1/2}$. In particular, if X_0 is a compact metric graph, then $\|D_\varepsilon\| = \mathrm{O}(\varepsilon^{1/2})$ without any assumption.

Note that the operator norm of A_ε in $D_\varepsilon = A^*_\varepsilon A_0 + B^*_\varepsilon B_0$ leads to the error estimate $\mathrm{O}(\varepsilon^{1/2})$, as it is dominant if $\varepsilon \to 0$.

5. Generalized Norm Resolvent Convergence

Let L_ε be a family of self-adjoint and non-negative operators ($\varepsilon \ge 0$) acting in an ε-independent Hilbert space $\mathscr{H}$. We say that L_ε converges in the *norm resolvent sense* to L_0 if:

$$\|(L_\varepsilon+1)^{-1} - (L_0+1)^{-1}\|_{\mathscr{H}\to\mathscr{H}} \to 0.$$

As a consequence, operator functions of L_ε also converge in the norm, e.g., for the semigroups, we have:

$$\|\mathrm{e}^{-tL_\varepsilon} - \mathrm{e}^{-tL_0}\|_{\mathscr{H}\to\mathscr{H}} \to 0.$$

Moreover, the spectra converge uniformly on bounded intervals. In particular, if L_ε all have a purely discrete spectrum, then $\lambda_k(L_\varepsilon) \to \lambda_k(L_0)$, where $\lambda_k(\cdot)$ denotes the kth eigenvalue ordered increasingly and repeated with respect to multiplicity.

We now want to extend these results to operators acting in *different* Hilbert spaces.

Definition 1. *For $\varepsilon \ge 0$, let L_ε be a self-adjoint and non-negative operator acting in a Hilbert space $\mathscr{H}_\varepsilon$. We say that L_ε converges to L_0 in the generalized norm resolvent sense, if there is a family of bounded operators $J_\varepsilon \colon \mathscr{H}_0 \longrightarrow \mathscr{H}_\varepsilon$ such that:*

$$\|R_\varepsilon J_\varepsilon - J_\varepsilon R_0\|_{\mathscr{H}_0\to\mathscr{H}_\varepsilon} \to 0, \qquad J^*_\varepsilon J_\varepsilon = \mathrm{id}_{\mathscr{H}_0} \quad \text{and} \quad \|(\mathrm{id}_{\mathscr{H}_\varepsilon} - J_\varepsilon J^*_\varepsilon)R_\varepsilon\|_{\mathscr{H}_\varepsilon\to\mathscr{H}_\varepsilon} \to 0, \tag{10}$$

where $R_\varepsilon := (L_\varepsilon+1)^{-1}$ denotes the resolvent.

There are actually more general versions of generalized norm resolvent convergence; see, e.g., [2,3] or also [4] and the references therein. We can also specify the convergence speed as the maximum of the two norm estimates.

Moreover, almost all conclusions that hold for norm resolvent convergence are still true here, e.g., the convergence of eigenvalues or the spectrum. Moreover, if L_ε converges to L_0 in the generalized norm resolvent sense with convergence speed $\delta_\varepsilon \to 0$, then the corresponding semigroups converge, i.e., we have, e.g.,

$$\|\mathrm{e}^{-tL_\varepsilon} - J_\varepsilon \mathrm{e}^{-tL_0} J^*_\varepsilon\|_{\mathscr{H}_\varepsilon\to\mathscr{H}_\varepsilon} \le C_t\delta_\varepsilon \to 0, \qquad \varepsilon \to 0.$$

One can even control the dependency on t ($C_t = \mathrm{O}(1/t)$ as $t \to 0$); see ([4], Ex. 1.10 (ii)) for details.

As an application, we show that the corresponding solutions of the heat equations converge: denote by u_t, resp. f_t, the solution of

$$\partial f_t + L_0 f_t = 0 \qquad \text{and} \qquad \partial u_t + L_\varepsilon u_t = 0$$

with initial data $f_0 = J_\varepsilon^* u_0$ at $t = 0$, then we have:

$$\|u_t - J_\varepsilon f_t\|_{\mathscr{H}_\varepsilon} = \|(e^{-tL_\varepsilon} - J_\varepsilon e^{-tL_0} J_\varepsilon^*)u_0\|_{\mathscr{H}_\varepsilon} \leq C_t \delta_\varepsilon \|u_0\|_{\mathscr{H}_\varepsilon}, \tag{11}$$

i.e., the approximate solution $J_\varepsilon f_t$ converges to the proper solution u_ε of the more complicated problem on $\mathscr{H}_\varepsilon$ uniformly with respect to the initial data u_0.

We have already shown the first norm convergence and the equality in (10) in the previous section (cf. Theorem 1); but we even have:

Theorem 2. *Under the uniformity assumptions (2) and (8), the Neumann Laplacians L_ε on the graph-like space X_ε converge to the Kirchhoff Laplacian on the underlying metric graph X_0 in the generalized norm resolvent sense.*

Proof. It remains to show the last limit in (10). We have:

$$\|u - J_\varepsilon J_\varepsilon^* u\|^2_{\mathscr{H}_\varepsilon} = \sum_{v \in V} \|u\|^2_{L_2(X_{\varepsilon,v})} + \varepsilon^m \sum_{e \in E} \int_{I_e} \|u_e(s,\cdot) - \int_B u_e(s,y)\, dy\|^2_{L_2(B)}\, ds.$$

The integrand in the second sum can be estimated by:

$$\|u_e(s,\cdot) - \int_B u_e(s,y)\, dy\|^2_{L_2(B)} \leq \frac{1}{\lambda_2(B)} \|\nabla_B u(s,\cdot)\|^2_{L_2(B)}$$

using again the variational characterization of eigenvalues. In particular, the second sum can be estimated by $\varepsilon \mathfrak{l}_\varepsilon(u)$. The first sum is also small, as functions with bounded energy do not concentrate at the vertex neighborhoods $X_{\varepsilon,v}$. The arguments to show this (actually, $\|u\|^2_{L_2(X_{\varepsilon,v})} \leq O(\varepsilon)\mathfrak{l}_\varepsilon(u)$) are very similar to the ones used in the proof of Lemma 2. Details can be found, e.g., in ([3], Section 6.3). □

Note that, once having proven the generalized norm resolvent convergence, with an error term of order $\varepsilon^{1/2}$, we can approximately solve the heat equation on X_ε as in (11): note that on a metric graph, one might even find explicit formulas for the solutions of the heat equation f_t, at least for simple metric graphs; hence, one has automatically approximate solutions for the corresponding heat equation on the more complicated space X_ε.

Let us now come back to the original thick graph given by $X_\varepsilon^\diamond$, where the edge neighborhoods have slightly shorter edge lengths.

We say that two operators L_ε and $L_\varepsilon^\diamond$ are *asymptotically close in the generalized norm resolvent sense*, if (10) holds with $R_\varepsilon = (L_\varepsilon + 1)^{-1}$ and R_0 replaced by $(L_\varepsilon^\diamond + 1)^{-1}$. We have the following result (for the proof, see, e.g., ([3], Prp. 4.2.5):

Lemma 3. *If L_ε converges to L_0 and if L_ε and $L_\varepsilon^\diamond$ are asymptotically close, both in the generalized norm resolvent sense, then $L_\varepsilon^\diamond$ converges to L_0 in the generalized norm resolvent sense.*

Now, in our concrete example with the slightly shortened edges, we have (for a proof, see ([3], Prp. 5.3.7)):

Lemma 4. *Assume that L_ε and $L_\varepsilon^\diamond$ are given as in Section 3, then L_ε and $L_\varepsilon^\diamond$ are asymptotically close in the generalized norm resolvent sense.*

We then immediately conclude from Theorem 2:

Corollary 1. *Under the uniformity assumptions* (2) *and* (8)*, the Neumann Laplacians* $L^\diamond_\varepsilon$ *on the* ε/ω_m*-neighborhood* $X^\diamond_\varepsilon$ *of an embedded metric graph* $X_0 \subset \mathbb{R}^{m+1}$ *converge to the Kirchhoff Laplacian on* X_0 *in the generalized norm resolvent sense.*

6. Outlook

The author is currently working on extending this result to some mildly non-linear equations with Claudio Cacciapuoti and with Michael Hinz and Jan Simmer in two different settings. Probably, the first systematic treatment of (non-linear) partial differential operators on thin domains was given in the nice overview of Geneviéve Raugel [8], combining some abstract results with concrete examples, but to the best of our knowledge, no thick graph domain and its limit were considered there explicitly. For Neumann Laplacians on thick graphs, there were actually results about the convergence of certain non-linear problems in [9,10], but Kosugi's papers did not contain an abstract approach using identification operators as we do.

At the conference, Jean-Guy Caputo also presented results on non-linear waves in networks and thick graphs justifying at least numerically the Kirchhoff vertex conditions; see [11,12]. There is another interesting application of the concept of generalized norm resolvent convergence: Berkolaiko et al. [13] studied the behavior of Laplacians on metric graphs if some edge lengths shrink to zero. A similar result (a compact part of the metric graph shrinks to a point) using different methods has been presented by Cacciapuoti [14] at the conference. A general convergence scheme also for some mildly non-linear equations would allow extending their analysis to non-linear problems.

We have the following type of equations in mind. Let:

$$\partial_t u_t = L_\varepsilon u_t + F_\varepsilon(u_t),$$

for $\varepsilon > 0$ and

$$\partial_t f_t = L_0 f_t + F_0(f_t).$$

As the non-linearity, we think of $F_\varepsilon(\psi) = \alpha_\varepsilon |\psi|^{2\mu}\psi$ for some $\mu > 0$ and $\alpha_\varepsilon > 0$. For the the solution, we make the ansatz:

$$u_t = \mathrm{e}^{-tL_\varepsilon} u_0 - \int_0^t \mathrm{e}^{-(t-s)} L_\varepsilon F_\varepsilon(u_s)\,\mathrm{d}s$$

and similarly for f_t. The non-linearity and the identification operators have to fulfil some compatibility conditions, namely $F_\varepsilon \circ J_\varepsilon - J_\varepsilon \circ F_0$ has to be small in some sense. One might use an iteration procedure in order to obtain a sequence of functions converging to the solution. If $F_\varepsilon(\psi) = \alpha_\varepsilon |\psi|^{2\mu}\psi$ in our example of thick metric graphs converging to metric graphs, then we must have $\alpha_\varepsilon = \varepsilon^{m\mu}\alpha_0$.

If one wants to consider the non-linear Schrödinger equation $\mathrm{i}\partial_t u_t = L_\varepsilon u_t + F_\varepsilon(u_t)$, one faces the additional problem that the (generalized) norm resolvent convergence does not imply norm convergence of the unitary group $\mathrm{e}^{\mathrm{i}tL_\varepsilon}$ for general initial data u_0; if one restricts u_0 to the range of the spectral projection $\mathbb{1}[0,\lambda_0](L_\varepsilon)$ for some $\lambda_0 > 0$, then there are still some operator norm estimates; see ([3], Thm. 4.2.16) for details. Nevertheless, one also has to make sure that $F_\varepsilon(u_0)$ still remains in the range of $\mathbb{1}[0,\lambda_0](L_\varepsilon)$, which is probably too restrictive.

Funding: This research received no external funding.

Acknowledgments: O.P. would like to thank the organizers of the workshop "Nonlinear PDEs on Metric Graphs and Branched Networks" in Leiden/NL in August 2018 for the kind invitation. O.P. would also like to thank the organizers and the staff of the Lorentz Center for providing a beautiful atmosphere for the workshop. The author would also like to thank the anonymous referee for pointing out the nice overview article [8].

Conflicts of Interest: The author declares no conflict of interest.

References

1. Berkolaiko, G.; Kuchment, P. Introduction to quantum graphs. In *Mathematical Surveys and Monographs*; American Mathematical Society: Providence, RI, USA, 2013.
2. Post, O. Spectral convergence of quasi-one-dimensional spaces. *Ann. Henri Poincaré* **2006**, *7*, 933–973. [CrossRef]
3. Post, O. Spectral analysis on graph-like spaces. In *Lecture Notes in Mathematics*; Springer: Berlin, Germany, 2012, doi:10.1007/978-3-642-23840-6.
4. Post, O.; Simmer, J. Quasi-unitary equivalence and generalized norm resolvent convergence. *Rev. Roumaine Math. Pures Appl.* **2019**, *64*, 2–3.
5. Exner, P.; Post, O. Convergence of spectra of graph-like thin manifolds. *J. Geom. Phys.* **2005**, *54*, 77–115. [CrossRef]
6. Kuchment, P.; Zeng, H. Convergence of spectra of mesoscopic systems collapsing onto a graph. *J. Math. Anal. Appl.* **2001**, *258*, 671–700. [CrossRef]
7. Rubinstein, J.; Schatzman, M. Variational problems on multiply connected thin strips. I. Basic estimates and convergence of the Laplacian spectrum. *Arch. Ration. Mech. Anal.* **2001**, *160*, 271–308. [CrossRef]
8. Raugel, G. Dynamics of partial differential equations on thin domains. In *Dynamical Systems (Montecatini Terme, 1994)*; Lecture Notes in Math; Springer: Berlin, Germany, 1995; Volume 1609, pp. 208–315. doi:10.1007/BFb0095241.
9. Kosugi, S. A semilinear elliptic equation in a thin network-shaped domain. *J. Math. Soc. Japan* **2000**, *52*, 673–697. doi:10.2969/jmsj/05230673. [CrossRef]
10. Kosugi, S. Semilinear elliptic equations on thin network-shaped domains with variable thickness. *J. Differ. Equ.* **2002**, *183*, 165–188. doi:10.1006/jdeq.2001.4119. [CrossRef]
11. Caputo, J.G.; Dutykh, D. Nonlinear waves in networks: A simple approach using the sine-Gordon equation. *Phys. Rev. E* **2014**, *90*, 022912. [CrossRef] [PubMed]
12. Caputo, J.G.; Dutykh, D.; Gleyse, D. Coupling conditions for the nonlinear shallow water equations on a network. *arXiv* **2015**, arXiv:1509.09082.
13. Berkolaiko, G.; Latushkin, Y.; Sukhtaiev, S. Limits of quantum graph operators with shrinking edges. *arXiv* **2018**, arXiv:1806.00561 .
14. Cacciapuoti, C. Scale invariant effective Hamiltonians for a graph with a small compact core *arXiv* **2019**, arXiv:1903.01898.

symmetry

MDPI

Article

Scale Invariant Effective Hamiltonians for a Graph with a Small Compact Core

Claudio Cacciapuoti

Dipartimento di Scienza e Alta Tecnologia, Sezione di Matematica, Università dell'Insubria, Via Valleggio 11, 22100 Como, Italy; claudio.cacciapuoti@uninsubria.it

Received: 15 January 2019; Accepted: 6 March 2019; Published: 9 March 2019

Abstract: We consider a compact metric graph of size ε and attach to it several edges (leads) of length of order one (or of infinite length). As ε goes to zero, the graph $\mathcal{G}^\varepsilon$ obtained in this way looks like the star-graph formed by the leads joined in a central vertex. On $\mathcal{G}^\varepsilon$ we define an Hamiltonian H^ε, properly scaled with the parameter ε. We prove that there exists a scale invariant effective Hamiltonian on the star-graph that approximates H^ε (in a suitable norm resolvent sense) as $\varepsilon \to 0$. The effective Hamiltonian depends on the spectral properties of an auxiliary ε-independent Hamiltonian defined on the compact graph obtained by setting $\varepsilon = 1$. If zero is not an eigenvalue of the auxiliary Hamiltonian, in the limit $\varepsilon \to 0$, the leads are decoupled.

Keywords: metric graphs; scaling limit; Kreĭn formula; point interactions

MSC: 81Q35; 47A10; 34B45

1. Introduction

One nice feature of quantum graphs (metric graphs equipped with differential operators) is that they are simple objects. In many cases, for example in the framework of the analysis of self-adjoint realizations of the Laplacian, it is possible to write down explicit formulae for the relevant quantities, such as the resolvent or the scattering matrix (see, e.g., [1] and [2]).

If the graph is too intricate though, it can be difficult, if not impossible, to perform exact computations. In such a situation, one may be interested in a simpler, effective model which captures only the most essential features of a complex quantum graph.

If several edges of the graph are much shorter then others, an effective model should rely on a simpler graph obtained by shrinking the short edges into vertices. These new vertices should keep track of at least some of the spectral or scattering properties of the shrinking edges, and perform as a black box approximation for a small, possibly intricate, network.

Our goal is to understand under what circumstances this type of effective models can be implemented. In this report we give some preliminary results showing that under certain assumptions such approximation is possible.

To fix the ideas, consider a compact metric graph $\mathcal{G}^{in,\varepsilon}$ of size (total length) ε, and attach to it several edges of length of order one (or of infinite length), the *leads*. Clearly, when ε goes to zero, the graph obtained in this way (let us denote it by $\mathcal{G}^\varepsilon$) looks like the star-graph formed by the leads joined in a central vertex. Let us denote by $\mathcal{G}^{out}$ such star-graph and by v_0 the central vertex.

Given a certain Hamiltonian (self-adjoint Schrödinger operator) H^ε on $\mathcal{G}^\varepsilon$, we want to show that there exists an Hamiltonian H^{out} on $\mathcal{G}^{out}$ such that, for small ε, H^{out} approximates (in a sense to be specified) H^ε. Of course, one main issue is to understand what boundary conditions in the vertex v_0 characterize the domain of H^{out}.

It turns out that, under several technical assumptions, the boundary conditions in v_0 are fully determined by the spectral properties of an auxiliary, ε-independent Hamiltonian defined on the graph $\mathcal{G}^{in} = \mathcal{G}^{in,\varepsilon=1}$.

Below we briefly discuss these technical assumptions, and refer to Section 2 for the details.

(i) The Hamiltonian H^ε on $\mathcal{G}^\varepsilon$ is a self-adjoint realization of the operator $-\Delta + B^\varepsilon$ on $\mathcal{G}^\varepsilon$, where B^ε is a potential term.

(ii) To set up the graph $\mathcal{G}^\varepsilon$ we select N distinct vertices in $\mathcal{G}^{in,\varepsilon}$ (we call them *connecting vertices*) and attach to each of them one lead, which is either a finite or an infinite length edge. The domain of H^ε is characterized by *Kirchhoff* (also called *standard* or *free*) boundary conditions at the connecting vertices, i.e., in each connecting vertex functions are continuous and the sum of the outgoing derivatives equals zero.

(iii) Scale invariance; the small (or *inner*) part of the graph scales uniformly in ε, i.e., $\mathcal{G}^{in,\varepsilon} = \varepsilon\mathcal{G}^{in}$. The Hamiltonian H^ε has a specific scaling property with respect to the parameter ε; loosely speaking, up to a multiplicative factor, the "restriction" of H^ε to $\mathcal{G}^{in,\varepsilon}$ is unitarily equivalent to an ε-independent operator on $\mathcal{G}^{in}$. The scale invariance property can be made precise by reasoning in terms of Hamiltonians on the inner graph $\mathcal{G}^{in,\varepsilon}$. This is done in Section 4 below. Here we just mention that this assumption forces the scaling on the *in* component of the potential $B^{in,\varepsilon}(x) = \varepsilon^{-2}B^{in}(x/\varepsilon)$, $x \in \mathcal{G}^{in,\varepsilon}$, and, in the vertices of $\mathcal{G}^{in,\varepsilon}$, the Robin-type vertex conditions (if any) also scale with ε accordingly.

(iv) The "restriction" of H^ε to the leads does not depend on ε. In particular, B^{out}, the *out* component of the potential, does not depend on ε.

We prove that it is always possible to identify an Hamiltonian H^{out} on $\mathcal{G}^{out}$ that approximates the Hamiltonian H^ε. The Hamiltonian H^{out} is a self-adjoint realization of the operator $-\Delta + B^{out}$ on $\mathcal{G}^{out}$, and it is characterized by scale invariant vertex conditions in v_0, i.e., vertex conditions with no Robin part (see [3], Section 1.4.2); in our notation, scale invariant means $\Theta_v = 0$ in Equation (1). The precise form of the possible effective Hamiltonians is given in Definitions 6 and 7 below.

The convergence of H^ε to H^{out} is understood in the following sense. We look at the resolvent operator $R^\varepsilon_z := (H^\varepsilon - z)^{-1}$, $z \in \mathbb{C}\backslash\mathbb{R}$, as an operator in the Hilbert space $L^2(\mathcal{G}^\varepsilon) = L^2(\mathcal{G}^{out}) \oplus L^2(\mathcal{G}^{in,\varepsilon})$. In the limit $\varepsilon \to 0$, the bounded operator R^ε_z converges to an operator which is diagonal in the decomposition $L^2(\mathcal{G}^{out}) \oplus L^2(\mathcal{G}^{in,\varepsilon})$. The *out/out* component of the limiting operator is the resolvent of a self-adjoint operator in $L^2(\mathcal{G}^{out})$, which we identify as the effective Hamiltonian on the star-graph.

Additionally, we characterize the limiting boundary conditions in the vertex v_0 in terms of the spectral properties of an auxiliary Hamiltonian on the (compact) graph $\mathcal{G}^{in} = \mathcal{G}^{in,\varepsilon=1}$. We distinguish two mutually exclusive cases: in one case (that we call *generic*) zero is not an eigenvalue of the auxiliary Hamiltonian; in the other case (we call it *non-generic*) zero is an eigenvalue of the auxiliary Hamiltonian.

In the generic case the effective Hamiltonian, denoted by $\mathring{H}^{out}$, is characterized by *Dirichlet* (also called *decoupling*) boundary conditions in the vertex v_0, i.e., functions in its domain are zero in v_0, see Definition 6. From the point of view of applications this is the less interesting case, since the leads are decoupled (no transmission through v_0 is possible).

In the non-generic case the situation is more involved. If zero is an eigenvalue of the auxiliary Hamiltonian one can identify a corresponding set of orthonormal eigenfunctions (in general eigenvalues can have multiplicity larger than one, included the zero eigenvalue). In the domain of the effective Hamiltonian $\widehat{H}^{out}$, the boundary conditions in v_0 are associated to the values of these eigenfunctions in the connecting vertices, see Definition 7. In this case, the boundary conditions in the vertex v_0 are scale invariant but, in general, not of decoupling type. For example, if the multiplicity of the zero eigenvalue is one, and the corresponding eigenfunction assumes the same value in all the connecting vertices, the boundary conditions are of Kirchhoff type.

The proof of the convergence is based on a Kreĭn-type formula for the resolvent R^ε_z. This formula allows us to write R^ε_z as a block matrix operator in the decomposition $L^2(\mathcal{G}^\varepsilon) = L^2(\mathcal{G}^{out}) \oplus L^2(\mathcal{G}^{in,\varepsilon})$ (see Equation (31)). In the formula, the first term, $\mathring{R}^\varepsilon_z$, is block diagonal and contains the resolvents of $\mathring{H}^{out}$ and $\mathring{H}^{in,\varepsilon}$ (a scaled down version of the auxiliary Hamiltonian, see Section 2.4); the second term is non-trivial, and couples the *out* and *in* components to reconstruct the resolvent of the full Hamiltonian H^{ε}. As ε goes to zero, the off-diagonal components in R^{ε}_{z} converge to zero, hence, the *out* and *in* components are always decoupled in the limit. A careful analysis of the non-trivial term in Formula (31) shows that it converges to zero in the generic case. In the non-generic case, instead, the *out*/*out* component of the non-trivial term converges to a finite operator, and the whole *out*/*out* component of R^{ε}_{z} reconstructs the resolvent of the effective Hamiltonian $\widehat{H}_0$.

The limiting behavior of H^{ε} is essentially determined by the small ε asymptotics of the spectrum of the inner Hamiltonian $\mathring{H}^{in,\varepsilon}$. The scale invariance assumption implies that the eigenvalues of $\mathring{H}^{in,\varepsilon}$ are given by $\lambda^{\varepsilon}_{n} = \lambda_n/\varepsilon^2$, where λ_n are the eigenvalues of the (scaled up) auxiliary Hamiltonian $\mathring{H}^{in}$. Obviously, all the non-zero eigenvalues move to infinity as $\varepsilon \to 0$; the zero eigenvalue instead, if it exists, persists, and for this reason it plays a special role in the analysis.

Closely related to our work is the paper by G. Berkolaiko, Y. Latushkin, and S. Sukhtaiev [4], to which we refer also for additional references. In [4] the authors analyze the convergence of Schrödinger operators on metric graphs with shrinking edges. Our setting is similar to the one in [4] with several differences. In [4] there are no restrictions on the topology of the graph, i.e., $\mathcal{G}^{out}$ is not necessarily a star-graph; outer edges can form loops, be connected among them or to arbitrarily intricate finite length graphs. In [4], moreover, the scale invariance assumption is missing. With respect to our work, however, the potential terms in [4] do not play an essential role in the limiting problem (because they are uniformly bounded in the scaling parameter).

As it was done in [4], to analyze the convergence of H^{ε} to H^{out}, since they are operators on different Hilbert spaces, one could use the notion of δ^{ε}-quasi unitary equivalence (or generalized norm resolvent convergence) introduced by P. Exner and O. Post in the series of works [5–9]. In Theorems 1 and 2 we state our main results in terms of the expansion of the resolvent in the decomposition $L^2(\mathcal{G}^{\varepsilon}) = L^2(\mathcal{G}^{out}) \oplus L^2(\mathcal{G}^{in,\varepsilon})$; and comment on the δ^{ε}-quasi unitary equivalence of the operators H^{ε} and $\mathring{H}^{out}$ (or $\widehat{H}^{out}$) in Remark 6.

Our analysis, with the scaling on the potential $B^{in,\varepsilon}(x) = \varepsilon^{-2}B^{in}(x/\varepsilon)$, is also related to the problem of approximating point-interactions on the real line through scaled potentials in the presence of a zero energy resonance, see, e.g., [10]. The same type of scaling arises naturally also in the study of the convergence of Schrödinger operators in thin waveguides to operators on graphs, see, e.g., [11–14].

Problems on graphs with a small compact core have been studied in several papers in the case in which $\mathcal{G}^{\varepsilon}$ is itself a star-graph, see, e.g., [15–19]. In particular, in the latter series of works, the authors point out the role of the zero energy eigenvalue.

Also related to our work is the problem of the approximation of vertex conditions through "physical Hamiltonians". In [20] (see also references therein), it is shown that all the possible self-adjoint boundary conditions at the central vertex of a star-graph, can be obtained as the limit of Hamiltonians with δ-interactions and magnetic field terms on a graph with a shrinking inner part.

Instead of looking at the convergence of the resolvent, a different approach consists in the analysis of the time dependent problem. This is done, e.g., in [21], for a tadpole-graph as the circle shrinks to a point.

The paper is structured as follows. In Section 2 we introduce some notation, our assumptions and present the main results, see Theorems 1 and 2. In Section 3 we discuss the Kreĭn formulae for the resolvents of H^{ε} and $\widehat{H}^{out}$ (the limiting Hamiltonian in the non-generic case). These formulae are the main tools in our analysis. In Section 4 we discuss the scale invariance properties of the auxiliary Hamiltonian, and other relevant operators. In Section 5 we prove Theorems 1 and 2. In doing so we present the results with a finer estimate of the remainder, see Theorems 3 and 4. We conclude the paper with two appendices: in Appendix A we briefly discuss the proofs of the Kreĭn resolvent formulae from Section 3; in Appendix B we prove some useful bounds on the eigenvalues and eigenfunctions of $\mathring{H}^{in}$.

Index of Notation

For the convenience of the reader we recall here the notation for the Hamiltonians used in our analysis. For the definitions we refer to Section 2 below.

- H^ε is the full Hamiltonian.
- $\mathring{H}^{in}$ is the auxiliary Hamiltonian
- $\mathring{H}^{in,\varepsilon}$ is the scaled down auxiliary Hamiltonian (see Definition 2 and Section 4); $\mathring{H}^{in} = \mathring{H}^{in,\varepsilon=1}$.
- $\mathring{H}^{out}$ is the effective Hamiltonian in the generic case.
- $\widehat{H}^{out}$ is the effective Hamiltonian in the non-generic case.
- $\mathring{H}^\varepsilon$ is the diagonal Hamiltonian $\mathring{H}^\varepsilon = \text{diag}(\mathring{H}^{out}, \mathring{H}^{in,\varepsilon})$ in the decomposition $L^2(\mathcal{G}^\varepsilon) = L^2(\mathcal{G}^{out}) \oplus L^2(\mathcal{G}^{in,\varepsilon})$ (see Section 3).

2. Preliminaries and Main Results

For a general introduction to metric graphs we refer to the monograph [3]. Here, for the convenience of the reader, we introduce some notation and recall few basic notions that will be used throughout the paper.

2.1. Basic Notions and Notation

To fix the ideas we start by selecting a collection of points, the vertices of the graph, and a connection rule among them. The bonds joining the vertices are associated to oriented segments and are the finite-length edges of the graph. Other edges can be of infinite length, and these edges are connected only to one vertex and are associated to half-lines. In this way we obtained a metric graph, see, e.g., Figure 1.

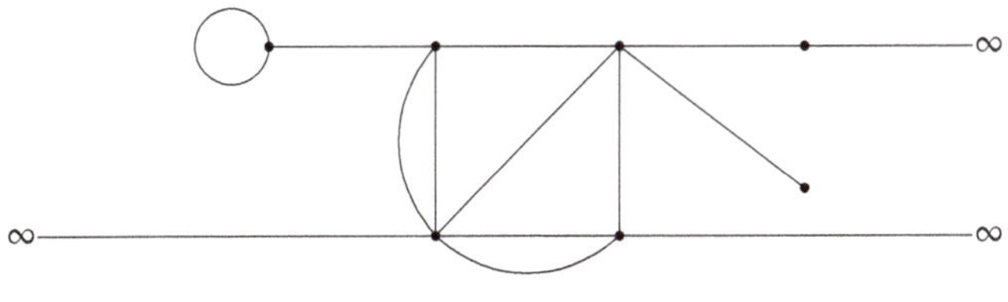

Figure 1. A metric graph with seven vertices (marked by dots) and 14 edges (three of which are half-lines).

Given a metric graph $\mathcal{G}$ we denote by $\mathcal{E}$ the set of its edges and by $\mathcal{V}$ the set of its vertices. We shall also use the notation $|\mathcal{E}|$ and $|\mathcal{V}|$ to denote the cardinality of $\mathcal{E}$ and $\mathcal{V}$ respectively. We shall always assume that both $|\mathcal{E}|$ and $|\mathcal{V}|$ are finite.

For any $e \in \mathcal{E}$, we identify the corresponding edge with the segment $[0, \ell_e]$ if e has finite length $\ell_e > 0$, or with $[0, +\infty)$ if e has infinite length.

Given a function $\psi : \mathcal{G} \to \mathbb{C}$, for $e \in \mathcal{E}$, ψ_e denotes its restriction to the edge e. With this notation in mind one can define the Hilbert space

$$\mathcal{H} := \bigoplus_{e\in\mathcal{E}} L^2(e),$$

with scalar product and norm given by

$$(\phi, \psi)_\mathcal{H} := \sum_{e\in\mathcal{E}} (\phi_e, \psi_e)_{L^2(e)} \qquad \text{and} \qquad \|\psi\|_\mathcal{H} := (\psi, \psi)_\mathcal{H}^{1/2}.$$

In a similar way one can define the Sobolev space $\mathcal{H}_2 := \bigoplus_{e\in\mathcal{E}} H^2(e)$, equipped with the norm

$$\|\psi\|_{\mathcal{H}_2} := \left(\sum_{e\in\mathcal{E}} \|\psi_e\|^2_{H^2(e)}\right)^{1/2}.$$

Note that functions in $\mathcal{H}_2$ are continuous in the edges of the graph but do not need to be continuous in the vertices.

For any vertex $v \in \mathcal{V}$ we denote by $d(v)$ the degree of the vertex, this is the number of edges having one endpoint identified by v, counting twice the edges that have both endpoints coinciding with v (loops). Let $\mathcal{E}_v \subseteq \mathcal{E}$ be the set of edges which are incident to the vertex v. For any vertex v we order the edges in $\mathcal{E}_v$ in an arbitrary way, counting twice the loops. In this way, for an arbitrary function $\psi \in \mathcal{H}_2$, one can define the vector $\Psi(v) \in \mathbb{C}^{d(v)}$ associated to the evaluation of ψ in v, i.e., the components of $\Psi(v)$ are given by $\psi_e(0)$ or $\psi_e(\ell_e)$, $e \in \mathcal{E}_v$, depending whether v is the initial or terminal vertex of the edge e, or by both values if e is a loop.

In a similar way one can define the vector $\Psi'(v) \in \mathbb{C}^{d(v)}$ with components $\psi_e'(0)$ and $-\psi_e'(\ell_e)$, $e \in \mathcal{E}_v$. Note that in the definition of $\Psi'(v)$, ψ_e' denotes the derivative of $\psi_e(x)$ with respect to x, and the derivative in v is always taken in the outgoing direction with respect to the vertex.

We are interested in defining self-adjoint operators in $\mathcal{H}$ which coincide with the Laplacian, possibly plus a potential term.

We denote by B the potential term in the operator, so that $B : \mathcal{G} \to \mathbb{R}$ is a real-valued function on the graph; and denote by B_e its restriction to the edge e. Additionally we assume that B is bounded and compactly supported on $\mathcal{G}$.

For every vertex $v \in \mathcal{V}$ we define a projection $P_v : \mathbb{C}^{d(v)} \to \mathbb{C}^{d(v)}$ and a self-adjoint operator Θ_v in Ran P_v, both P_v and Θ_v can be identified with Hermitian $d(v) \times d(v)$ matrices.

It is well known, see, e.g., [3] and ([22], Example 5.2) that the operator $H_{P,\Theta}$ defined by:

$$D(H_{P,\Theta}) := \left\{\psi \in \mathcal{H}_2 |\, P_v^\perp \Psi(v) = 0;\; P_v\Psi'(v) - \Theta_v P_v \Psi(v) = 0 \quad \forall v \in \mathcal{V}\right\} \tag{1}$$

$$(H_{P,\Theta}\psi)_e := -\psi_e'' + B_e\psi_e \qquad \forall e \in \mathcal{E} \tag{2}$$

is self-adjoint. Instead of Equation (2), we shall write

$$H_{P,\Theta}\psi := -\psi'' + B\psi, \tag{3}$$

to be understood componentwise.

We remark that for every P_v and Θ_v as above, $H_{P,\Theta}$ is a self-adjoint extension of the symmetric operator H_{min}

$$D(H_{min}) := \{\psi \in \mathcal{H}_2 |\, \Psi(v) = 0;\; \Psi'(v) = 0 \quad \forall v \in \mathcal{V}\} \qquad H_{min}\psi := -\psi'' + B\psi.$$

2.2. Graphs with a Small Compact Core

We consider a graph $\mathcal{G}^\varepsilon$ obtained by attaching several edges to a small compact core (a compact metric graph of size ε).

We denote the compact core of the graph by $\mathcal{G}^{in,\varepsilon}$. The graph $\mathcal{G}^{in,\varepsilon}$ is obtained by shrinking a compact graph $\mathcal{G}^{in}$ by means of a parameter $0 < \varepsilon < 1$, more precisely, we set

$$\mathcal{G}^{in,\varepsilon} = \varepsilon \mathcal{G}^{in}. \tag{4}$$

We denote by $\mathcal{E}^{in}$ the set of edges of the graph $\mathcal{G}^{in}$ and by $\mathcal{E}^{in,\varepsilon}$ the set of edges of the graph $\mathcal{G}^{in,\varepsilon}$.

In the graph $\mathcal{G}^{in}$ (or, equivalently, in $\mathcal{G}^{in,\varepsilon}$) we select N distinct vertices that we label with $v_1, ..., v_N$, and refer to them as connecting vertices. We shall denote by C the set of connecting vertices. We denote by $\mathcal{V}^{in}$ the set of all the remaining vertices, and call the elements of $\mathcal{V}^{in}$ *inner vertices* (note that the set $\mathcal{V}^{in}$ may be empty).

To construct the graph $\mathcal{G}^{\varepsilon}$, we attach to each connecting vertex one additional edge which can be an half-line or an edge of finite length (not dependent on ε). We shall call these additional edges *outer edges* and denote by $\mathcal{E}^{out}$ the corresponding set of edges; obviously $|\mathcal{E}^{out}| = N$. When needed, we shall denote these edges by $e_1, ..., e_N$, so that the edge e_j is connected to the vertex v_j, $j = 1, ..., N$. Moreover we shall use the notation

$$\psi_{e_j} \equiv \psi_j \qquad e_j \in \mathcal{E}^{out},\ j = 1, ..., N.$$

Note that if $e \in \mathcal{E}^{out}$ is of finite length the endpoint which does not coincide with the connecting vertex is of degree one (all the finite length outer edges are pendants).

We shall always assume, without loss of generality, that for each edge in $\mathcal{E}^{out}$ the connecting vertex is identified by $x = 0$.

We denote by $\mathcal{E}^{\varepsilon}$ and $\mathcal{V}$ the sets of edges and vertices of the graph $\mathcal{G}^{\varepsilon}$. We note that $\mathcal{E}^{\varepsilon} = \mathcal{E}^{out} \cup \mathcal{E}^{in,\varepsilon}$ and $\mathcal{V} = \mathcal{V}^{out} \cup \mathcal{C} \cup \mathcal{V}^{in}$, where $\mathcal{V}^{out}$ is the set of vertices in $\mathcal{G}^{\varepsilon}$ which are neither connecting nor inner vertices.

Remark 1. *For any $v \in \mathcal{C}$ we denote by $d^{in}(v)$ its degree as a vertex of the graph $\mathcal{G}^{in,\varepsilon}$, so that its degree as a vertex of the graph $\mathcal{G}^{\varepsilon}$ is $d(v) = d^{in}(v) + 1$.*

As $\varepsilon \to 0$, the inner graph shrinks to one point, in the limit all the connecting vertices merge in one vertex which we identify with the point $x_j = 0$, x_j being the coordinate along the edge $e_j \in \mathcal{E}^{out}$, $j = 1, \ldots, N$. In the limit the graph $\mathcal{G}^{\varepsilon}$ looks like a star-graph with N edges connected in the origin, see Figure 2; we denote the star-graph by $\mathcal{G}^{out}$.

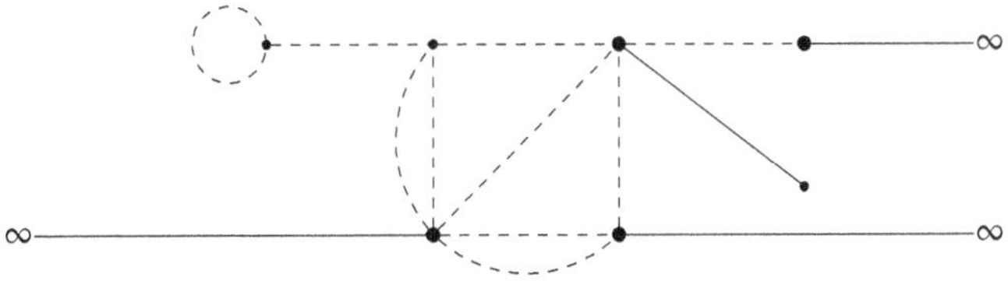

Figure 2. The dashed lines represent the edges of $\mathcal{G}^{in,\varepsilon}$, the large dots the connecting vertices. The graph $\mathcal{G}^{out}$ is obtained by merging the connecting vertices. In the example in the picture, $\mathcal{G}^{out}$ has three infinite edges and one edge of finite length.

We define the Hilbert spaces:

$$\mathcal{H}^{\varepsilon} := \bigoplus_{e \in \mathcal{E}^{\varepsilon}} L^2(e), \qquad \mathcal{H}^{out} := \bigoplus_{e \in \mathcal{E}^{out}} L^2(e), \qquad \mathcal{H}^{in,\varepsilon} := \bigoplus_{e \in \mathcal{E}^{in,\varepsilon}} L^2(e).$$

We remark that one can always think of $\mathcal{H}^{\varepsilon}$ as the direct sum

$$\mathcal{H}^{\varepsilon} = \mathcal{H}^{out} \oplus \mathcal{H}^{in,\varepsilon}, \tag{5}$$

and decompose each function $\psi \in \mathcal{H}^{\varepsilon}$ as $\psi = (\psi^{out}, \psi^{in})$ with $\psi^{out} \in \mathcal{H}^{out}$ and $\psi^{in} \in \mathcal{H}^{in,\varepsilon}$. When no misunderstanding is possible, we omit the dependence on ε, moreover we simply write ψ, instead of ψ^{out} or ψ^{in}.

In a similar way we introduce the Sobolev spaces

$$\mathcal{H}_2^{\varepsilon} := \bigoplus_{e \in \mathcal{E}^{\varepsilon}} H^2(e), \qquad \mathcal{H}_2^{out} := \bigoplus_{e \in \mathcal{E}^{out}} H^2(e), \qquad \mathcal{H}_2^{in,\varepsilon} := \bigoplus_{e \in \mathcal{E}^{in,\varepsilon}} H^2(e).$$

2.3. Full Hamiltonian

Next we define an Hamiltonian H^ε in $\mathcal{H}^\varepsilon$ (of the form given in Equations (1)–(3)); this is the object of our investigation.

- Recall that if $v \in \mathcal{V}^{out}$, then $d(v) = 1$. For any $v \in \mathcal{V}^{out}$ we fix an orthogonal projection $P_v^{out} : \mathbb{C} \to \mathbb{C}$, and a self-adjoint operator Θ_v^{out} in $\mathrm{Ran}(P_v^{out})$. Since vertices in $\mathcal{V}^{out}$ have degree one, P_v^{out} is either 1 or 0; whenever $P_v^{out} = 1$ it makes sense to define Θ_v^{out} which turns out to be the operator acting as the multiplication by a real constant. In other words, the boundary conditions in $v \in \mathcal{V}^{out}$ (of the form given in the definition of $D(H_{P,\Theta})$) can be of Dirichlet type, $\psi_e(v) = 0$; of Neumann type $\psi_e'(v) = 0$; or of Robin type $\psi_e'(v) = \alpha\psi_e(v)$ with $\alpha \in \mathbb{R}$.
It would be possible to consider a more general setting in which the outer graph has a non trivial topology, in same spirit of the work [4], but we will not pursue this goal.
- For any $v \in \mathcal{C}$ we define the orthogonal projection (see Remark 1 for the definition of $d(v)$):

$$K_v : \mathbb{C}^{d(v)} \to \mathbb{C}^{d(v)}, \qquad K_v := \mathbf{1}_{d(v)} \left(\mathbf{1}_{d(v)}, \cdot\right)_{\mathbb{C}^{d(v)}} \qquad \forall v \in \mathcal{C},$$

where $\mathbf{1}_{d(v)}$ denotes the vector (of unit norm) in $\mathbb{C}^{d(v)}$ defined by $\mathbf{1}_{d(v)} = (d(v))^{-1/2}(1, ..., 1)$. In a similar way, we define the orthogonal projection

$$K_v^{in} : \mathbb{C}^{d^{in}(v)} \to \mathbb{C}^{d^{in}(v)}, \qquad K_v^{in} := \mathbf{1}_{d^{in}(v)} \left(\mathbf{1}_{d^{in}(v)}, \cdot\right)_{\mathbb{C}^{d^{in}(v)}} \qquad \forall v \in \mathcal{C},$$

where $\mathbf{1}_{d^{in}(v)} \in \mathbb{C}^{d^{in}(v)}$ is defined by $\mathbf{1}_{d^{in}(v)} = (d^{in}(v))^{-1/2}(1, ..., 1)$. Both K_v and K_v^{in} have one-dimensional range given by the span of the vectors $\mathbf{1}_{d(v)}$ and $\mathbf{1}_{d^{in}(v)}$ respectively.
A function ψ satisfies Kirchhoff conditions in the vertex v (it is continuous in v and the sum of the outgoing derivatives in v equals zero) if and only if $K_v^\perp \Psi(v) = 0$ and $K_v \Psi'(v) = 0$.
- For any $v \in \mathcal{V}^{in}$ we fix an orthogonal projection $P_v^{in} : \mathbb{C}^{d(v)} \to \mathbb{C}^{d(v)}$, and a self-adjoint operator $\Theta_v^{in,\varepsilon}$ in $\mathrm{Ran}(P_v^{in})$.
- We fix an ε-dependent real-valued function $B^\varepsilon : \mathcal{G}^\varepsilon \to \mathbb{R}$, such that in the *out*/*in* decomposition (5) one has $B^\varepsilon = (B^{out}, B^{in,\varepsilon})$. With $B^{out} : \mathcal{G}^{out} \to \mathbb{R}$ bounded and compactly supported.
- Scale invariance; recall that $\mathcal{G}^{in,\varepsilon} = \varepsilon\mathcal{G}^{in}$, see Equation (4). We assume additionally: that $B^{in,\varepsilon}(x) = \varepsilon^{-2}B^{in}(x/\varepsilon)$, where $B^{in} : \mathcal{G}^{in} \to \mathbb{R}$ is bounded; and that $\Theta_v^{in,\varepsilon} = \varepsilon^{-1}\Theta_v^{in}$, for all $v \in \mathcal{V}^{in}$. For a discussion on the meaning and the main consequences of these assumptions we refer to Section 4.

Definition 1 (Hamiltonian H^ε). *We denote by H^ε the self-adjoint operator in $\mathcal{H}^\varepsilon$ defined by*

$$\begin{aligned} D(H^\varepsilon) := \{\psi \in \mathcal{H}_2^\varepsilon |\, & P_v^{in\perp}\Psi(v) = 0\,,\ P_v^{in}\Psi'(v) - \Theta_v^{in,\varepsilon}P_v^{in}\Psi(v) = 0 \quad \forall v \in \mathcal{V}^{in}; \\ & P_v^{out\perp}\Psi(v) = 0\,,\ P_v^{out}\Psi'(v) - \Theta_v^{out}P_v^{out}\Psi(v) = 0 \quad \forall v \in \mathcal{V}^{out}; \\ & K_v^\perp\Psi(v) = 0\,,\ K_v\Psi'(v) = 0 \quad \forall v \in \mathcal{C}\} \end{aligned}$$

$$H^\varepsilon\psi := -\psi'' + B^\varepsilon\psi \qquad \forall\psi \in D(H^\varepsilon).$$

Remark 2. *In the out/in decomposition one has*

$$(H^\varepsilon\psi)^{out} = -\psi^{out''} + B^{out}\psi^{out}$$

$$(H^\varepsilon\psi)^{in} = -\psi^{in''} + B^{in,\varepsilon}\psi^{in}.$$

Note that the action of the outer component of H^ε does not depend on ε.

Remark 3. *By the definition of K_v, in each connecting vertex boundary conditions in $D(H^\varepsilon)$ are of Kirchhoff-type: the function ψ is continuous in $v \in \mathcal{C}$ and*

$$\sum_{e\sim v} \psi_e'(v) = 0 \qquad v \in \mathcal{C},$$

where the sum is taken on all the edges incident on v (counting loops twice) and the derivative is understood in the outgoing direction from the vertex.

2.4. Auxiliary Hamiltonian

We are interested in the limit of the operator H^ε as $\varepsilon \to 0$. We shall see that the limiting properties of H^ε are strongly related to spectral properties of the Hamiltonian $\mathring{H}^{in,\varepsilon}$:

Definition 2 (Auxiliary Hamiltonian, scaled down version).

$$\begin{aligned} D(\mathring{H}^{in,\varepsilon}) := \{\psi \in \mathcal{H}_2^{in,\varepsilon} |\, P_v^{in\perp}\Psi(v) = 0\,,\ P_v^{in}\Psi'(v) - \Theta_v^{in,\varepsilon} P_v^{in}\Psi(v) = 0 \quad \forall v \in \mathcal{V}^{in}; \\ K_v^{in\perp}\Psi(v) = 0\,,\ K_v^{in}\Psi'(v) = 0 \quad \forall v \in \mathcal{C}\} \end{aligned} \tag{6}$$

$$\mathring{H}^{in,\varepsilon}\psi := -\psi'' + B^{in,\varepsilon}\psi \qquad \forall \psi \in D(\mathring{H}^{in,\varepsilon}).$$

Let $\mathcal{H}^{in} = \mathcal{H}^{in,\varepsilon=1}$, and define the unitary scaling group

$$U^{in,\varepsilon} : \mathcal{H}^{in} \to \mathcal{H}^{in,\varepsilon}\,, \qquad (U^{in,\varepsilon}\psi^{in})(x) := \varepsilon^{-1/2}\psi^{in}(x/\varepsilon);$$

its inverse is

$$U^{in,\varepsilon^{-1}} : \mathcal{H}^{in,\varepsilon} \to \mathcal{H}^{in}\,, \qquad (U^{in,\varepsilon^{-1}}\psi^{in})(x) = \varepsilon^{1/2}\psi^{in}(\varepsilon x).$$

By the scaling properties $\Theta_v^{in,\varepsilon} = \varepsilon^{-1}\Theta_v^{in}$ and $B^{in,\varepsilon}(x/\varepsilon) = \varepsilon^{-2}B^{in}(x)$, one infers the unitary relation

$$\mathring{H}^{in,\varepsilon} = \varepsilon^{-2} U^{in,\varepsilon} \mathring{H}^{in} U^{in,\varepsilon^{-1}} \tag{7}$$

with $\mathring{H}^{in}$ defined on $\mathcal{H}^\varepsilon$ and given by $\mathring{H}^{in} = \mathring{H}^{in,\varepsilon=1}$. One consequence of Equation (7) is that the spectrum of $\mathring{H}^{in,\varepsilon}$ is related to the spectrum of $\mathring{H}^{in}$ by the relation $\sigma(\mathring{H}^{in,\varepsilon}) = \varepsilon^{-2}\sigma(\mathring{H}^{in})$ (see Section 4 for more comments on the implications of the scale invariance assumption). For this reason, we prefer to formulate the results in terms of the spectral properties of the ε-independent Hamiltonian $\mathring{H}^{in}$ instead of the spectral properties of $\mathring{H}^{in,\varepsilon}$.

Definition 3 (Auxiliary Hamiltonian $\mathring{H}^{in}$). *We call Auxiliary Hamiltonian the Hamiltonian $\mathring{H}^{in} = \mathring{H}^{in,\varepsilon=1}$ defined on $\mathcal{H}^{in}$.*

Letting $\mathcal{H}_2^{in} = \mathcal{H}_2^{in,\varepsilon=1}$, the domain and action of $\mathring{H}^{in}$ are given by

$$\begin{aligned} D(\mathring{H}^{in}) = \{\psi \in \mathcal{H}_2^{in} |\, P_v^{in\perp}\Psi(v) = 0\,,\ P_v^{in}\Psi'(v) - \Theta_v^{in} P_v^{in}\Psi(v) = 0 \quad \forall v \in \mathcal{V}^{in}; \\ K_v^{in\perp}\Psi(v) = 0\,,\ K_v^{in}\Psi'(v) = 0 \quad \forall v \in \mathcal{C}\} \end{aligned} \tag{8}$$

$$\mathring{H}^{in}\psi = -\psi'' + B^{in}\psi \qquad \forall \psi \in D(\mathring{H}^{in}).$$

The spectrum of $\mathring{H}^{in}$ consists of isolated eigenvalues of finite multiplicity, see, e.g., ([3], Theorem 3.1.1). For $n \in \mathbb{N}$, we denote by λ_n the eigenvalues of $\mathring{H}^{in}$ (counting multiplicity) and by $\{\varphi_n\}_{n\in\mathbb{N}}$ a corresponding set of orthonormal eigenfunctions.

Definition 4 (Generic/non-generic case). *In the analysis of the limit of H^ε we distinguish two cases:*

(1) *Generic (or non-resonant, or decoupling) case.* $\lambda = 0$ *is not an eigenvalue of the operator* $\mathring{H}^{in}$.
(2) *Non-generic (or resonant) case.* $\lambda = 0$ *is an eigenvalue of the operator* $\mathring{H}^{in}$.
In the non-generic case we denote by $\{\hat{\varphi}_k\}_{k=1,\dots,m}$ *a set of (orthonormal) eigenfunctions corresponding to the zero eigenvalue. By Equation (8), functions in* $D(\mathring{H}^{in})$ *are continuous in the connecting vertices (see also Remark 3). We denote by* $\hat{\varphi}_k(v)$, $v \in \mathcal{C}$, *the value of* $\hat{\varphi}_k$ *in* v, *and define the vectors*

$$\underline{\hat{c}}_k := (\hat{\varphi}_k(v_1), \dots, \hat{\varphi}_k(v_N)) \in \mathbb{C}^N, \qquad k = 1, \dots, m, \; v_j \in \mathcal{C}, \; j = 1, \dots, N. \tag{9}$$

Definition 5 ($\widehat{C} - \widehat{P}$)**.** *In the non-generic case, let* $\widehat{C}$ *be the operator*

$$\widehat{C} := \sum_{k=1}^{m} \underline{\hat{c}}_k (\underline{\hat{c}}_k, \cdot)_{\mathbb{C}^N} : \mathbb{C}^N \to \mathbb{C}^N.$$

$\widehat{C}$ *is a bounded self-adjoint operator (it is an* $N \times N$ *Hermitian matrix). Denote by* $\operatorname{Ran}\widehat{C} \subseteq \mathbb{C}^N$ *and* $\operatorname{Ker}\widehat{C} \subseteq \mathbb{C}^N$, *the range and the kernel of* $\widehat{C}$ *respectively. One has that the subspaces* $\operatorname{Ran}\widehat{C}$ *and* $\operatorname{Ker}\widehat{C}$ *are* $\widehat{C}$*-invariant. Moreover,* $\mathbb{C}^N = \operatorname{Ran}\widehat{C} \oplus \operatorname{Ker}\widehat{C}$. *In what follows we denote by* $\widehat{P}$ *the orthogonal projection (Riesz projection, see, e.g., ([23], Section I.2)) on* $\operatorname{Ran}(\widehat{C})$, *and by* $\widehat{P}^\perp = \mathbb{I}_N - \widehat{P}$ *the orthogonal projection on* $\operatorname{Ker}(\widehat{C})$.

Remark 4. *We note that* $\underline{q} \in \operatorname{Ker}\widehat{C}$ *if and only if* $(\underline{\hat{c}}_k, \underline{q})_{\mathbb{C}^N} = 0$ *for all* $k = 1, \dots, m$. *To see that this indeed the case, observe that if* $\underline{q} \in \operatorname{Ker}\widehat{C}$ *then it must be* $(\underline{q}, \widehat{C}\underline{q})_{\mathbb{C}^N} = 0$, *hence,* $\sum_{k=1}^m |(\underline{\hat{c}}_k, \underline{q})_{\mathbb{C}^N}|^2 = 0$, *which in turn implies* $(\underline{\hat{c}}_k, \underline{q})_{\mathbb{C}^N} = 0$ *for all* $k = 1, \dots, m$. *The other implication is trivial.*

Since $\widehat{P}^\perp \underline{\hat{c}}_k \in \operatorname{Ker}\widehat{C}$, *we infer* $0 = (\underline{\hat{c}}_k, \widehat{P}^\perp \underline{\hat{c}}_k)_{\mathbb{C}^N} = (\widehat{P}^\perp \underline{\hat{c}}_k, \widehat{P}^\perp \underline{\hat{c}}_k)_{\mathbb{C}^N} = \|\widehat{P}^\perp \underline{\hat{c}}_k\|^2_{\mathbb{C}^N}$ *for all* $k = 1, \dots, m$; *hence,* $\widehat{P}^\perp \underline{\hat{c}}_k = 0$, *or, equivalently,* $\underline{\hat{c}}_k \in \operatorname{Ran}(\widehat{C})$.

2.5. Effective Hamiltonians

We shall see that the definition of the limiting operator (effective Hamiltonian in $\mathcal{H}^{out}$) depends on presence of a zero eigenvalue for $\mathring{H}^{in}$ (the occurrence of the generic case vs. the non-generic case).

Recall that for $\psi \in \mathcal{H}^{out}$, we used ψ_j to denote the component of ψ on the edge e_j attached to the connecting vertex v_j. Moreover, we assumed that the vertex v_j is identified by $x = 0$. With this remark in mind, given a function $\psi \in \mathcal{H}_2^{out}$ we define the vectors

$$\Psi(\mathbf{0}) := (\psi_1(0), \dots, \psi_N(0))^T \in \mathbb{C}^N, \qquad \Psi'(\mathbf{0}) := (\psi_1'(0), \dots, \psi_N'(0))^T \in \mathbb{C}^N.$$

These correspond to $\Psi(v_0)$ and $\Psi'(v_0)$, as defined in Section 2.1, where v_0 is the central vertex of the star-graph $\mathcal{G}^{out}$.

In the limit $\varepsilon \to 0$, the connecting vertices in $\mathcal{G}^{in,\varepsilon}$ coincide, and can be identified with the vertex $v_0 \equiv \mathbf{0}$.

We distinguish two possible effective Hamiltonians in $\mathcal{H}^{out}$.

Definition 6 (Effective Hamiltonian, generic case)**.** *We denote by* $\mathring{H}^{out}$ *the self-adjoint operator in* $\mathcal{H}^{out}$ *defined by*

$$\begin{aligned} D(\mathring{H}^{out}) := \{\psi \in \mathcal{H}_2^{out} |\, P_v^{out\perp}\Psi(v) = 0,\; P_v^{out}\Psi'(v) - \Theta_v^{out} P_v^{out}\Psi(v) = 0 \quad \forall v \in \mathcal{V}^{out}; \\ \Psi(\mathbf{0}) = 0\} \end{aligned} \tag{10}$$

$$H^{out}\psi := -\psi'' + B^{out}\psi \qquad \forall \psi \in D(\mathring{H}^{out}).$$

Definition 7 (Effective Hamiltonian, non-generic case). *Let $\widehat{P}$ be the orthogonal projection given in Definition 5. We denote by $\widehat{H}^{out}$ the self-adjoint operator in $\mathcal{H}^{out}$ defined by*

$$D(\widehat{H}^{out}) := \{\psi \in \mathcal{H}_2^{out} |\, P_v^{out\perp}\Psi(v) = 0\,,\; P_v^{out}\Psi'(v) - \Theta_v^{out}P_v^{out}\Psi(v) = 0 \quad \forall v \in \mathcal{V}^{out};$$
$$\widehat{P}^{\perp}\Psi(\mathbf{0}) = 0\,,\; \widehat{P}\Psi'(\mathbf{0}) = 0\}$$

$$\widehat{H}^{out}\psi := -\psi'' + B^{out}\psi \qquad \forall\psi \in D(\widehat{H}^{out}).$$

The boundary conditions in $\mathbf{0}$ in the definitions of $D(\mathring{H}^{out})$ and $D(\widehat{H}^{out})$ are scale invariant (see ([3], Section 1.4.2)).

2.6. Main Results

In what follows C denotes a generic positive constant independent on ε. Given two Hilbert spaces X and Y, we denote by $\mathcal{B}(X,Y)$ (or simply by $\mathcal{B}(X)$ if $X = Y$) the space of bounded linear operators from X to Y, and by $\|\cdot\|_{\mathcal{B}(X,Y)}$ the corresponding norm. For any $a \in \mathbb{R}$, we use the notation $\mathcal{O}_{\mathcal{B}(X,Y)}(\varepsilon^a)$ to denote a generic operator from X to Y whose norm is bounded by $C\varepsilon^a$ for ε small enough.

Given a bounded operator A in $\mathcal{H}^\varepsilon$ we use the notation

$$A = \begin{pmatrix} A^{out,out} & A^{out,in} \\ A^{in,out} & A^{in,in} \end{pmatrix} \tag{11}$$

to describe its action in the *out*/*in* decomposition (5): here $A^{u,v} : \mathcal{H}^v \to \mathcal{H}^u$, $u, v = out, in$, are operators defined according to

$$\begin{aligned}(A\psi)^{out} &= A^{out,out}\psi^{out} + A^{out,in}\psi^{in} \\ (A\psi)^{in} &= A^{in,out}\psi^{out} + A^{in,in}\psi^{in}.\end{aligned} \tag{12}$$

Theorem 1. *Let $z \in \mathbb{C}\backslash\mathbb{R}$. In the generic case (see Definition 4)*

$$(H^\varepsilon - z)^{-1} = \begin{pmatrix} (\mathring{H}^{out} - z)^{-1} & \mathbb{O} \\ \mathbb{O} & \mathbb{O} \end{pmatrix} + \mathcal{O}_{\mathcal{B}(\mathcal{H}^\varepsilon)}(\varepsilon), \tag{13}$$

where the expansion has to be understood in the out/in decomposition (11).

Theorem 2. *Let $z \in \mathbb{C}\backslash\mathbb{R}$. In the non-generic case (see Definition 4), let $\widehat{C}_0$ be the restriction of $\widehat{C}$ to* Ran $\widehat{P}$.

(i) If Ker $\widehat{C} \subset \mathbb{C}^N$, *$\widehat{C}_0$ is invertible as an operator in $\widehat{P}\mathbb{C}^N$, and*

$$(H^\varepsilon - z)^{-1} = \begin{pmatrix} (\widehat{H}^{out} - z)^{-1} & \mathbb{O} \\ \mathbb{O} & -z^{-1}\sum_{k,k'=1}^{m}\left(\delta_{k,k'} - (\underline{\hat{c}}_k, \widehat{C}_0^{-1}\underline{\hat{c}}_{k'})_{\mathbb{C}^N}\right)\varphi_k^\varepsilon(\varphi_{k'}^\varepsilon, \cdot)_{\mathcal{H}^{in,\varepsilon}} \end{pmatrix} + \mathcal{O}_{\mathcal{B}(\mathcal{H}^\varepsilon)}(\varepsilon^{1/2}), \tag{14}$$

where the expansion has to be understood in the out/in decomposition (11).

(ii) If Ker $\widehat{C} = \mathbb{C}^N$, *then $\widehat{P} = 0$, and expansion (14) holds true with $\widehat{H}^{out} = \mathring{H}^{out}$, $(\underline{\hat{c}}_k, \widehat{C}_0^{-1}\underline{\hat{c}}_{k'})_{\mathbb{C}^N} = 0$ for all $k, k' = 1, \ldots, m$, and the error term changed in $\mathcal{O}_{\mathcal{B}(\mathcal{H}^\varepsilon)}(\varepsilon)$.*

(iii) If the vectors $\underline{\hat{c}}_k$, $k = 1, \ldots, m$, are linearly independent, then $\left(\delta_{k,k'} - (\underline{\hat{c}}_k, \widehat{C}_0^{-1}\underline{\hat{c}}_{k'})_{\mathbb{C}^N}\right) = 0$ for all $k, k' = 1, \ldots, m$, and

$$(H^\varepsilon - z)^{-1} = \begin{pmatrix} (\widehat{H}^{out} - z)^{-1} & \mathbb{O} \\ \mathbb{O} & \mathbb{O} \end{pmatrix} + \mathcal{O}_{\mathcal{B}(\mathcal{H}^\varepsilon)}(\varepsilon^{1/2}). \tag{15}$$

Remark 5. *Finer estimates on the remainders in Equations (13) and (14) are given in Theorems 3 and 4 below.*

Remark 6. *We recall and adapt to our setting the notion of δ^ε-quasi unitary equivalence of operators acting on different Hilbert spaces introduced by P. Exner and O. Post, see in particular ([7], Section 3.2) and ([9], Chapter 4). See also ([4], Section 5) for a discussion on the application of this approach to the analysis of operators on graphs with shrinking edges.*

Let J be the operator

$$J : \mathcal{H}^{out} \to \mathcal{H}^\varepsilon, \qquad J\psi^{out} = (\psi^{out}, 0) \quad \text{for all } \psi^{out} \in \mathcal{H}^{out},$$

where $(\psi^{out}, 0)$ is understood in the decomposition (5). Its adjoint J^* maps $\mathcal{H}^\varepsilon$ in $\mathcal{H}^{out}$, and is given by:

$$J^* : \mathcal{H}^\varepsilon \to \mathcal{H}^{out}, \qquad J^*\psi = \psi^{out} \quad \text{for all } \psi = (\psi^{out}, \psi^{in}) \in \mathcal{H}^\varepsilon.$$

Note that $J^*J = \mathbb{I}^{out}$, where $\mathbb{I}^{out}$ is the identity in $\mathcal{H}^{out}$.

The operator H^ε is δ^ε-quasi unitarily equivalent to a self-adjoint operator H^{out} in $\mathcal{H}^{out}$ if

$$\|(\mathbb{I} - JJ^*)(H^\varepsilon - z)^{-1}\|_{\mathcal{B}(\mathcal{H}^\varepsilon)} \leq C\delta^\varepsilon \quad \text{and} \quad \|J(H^{out} - z)^{-1} - (H^\varepsilon - z)^{-1}J\|_{\mathcal{B}(\mathcal{H}^{out}, \mathcal{H}^\varepsilon)} \leq C\delta^\varepsilon, \tag{16}$$

for some $z \in \mathbb{C}\backslash\mathbb{R}$.

Note that in the decomposition (12), one has

$$(\mathbb{I} - JJ^*)(H^\varepsilon - z)^{-1}\psi = ((H^\varepsilon - z)^{-1})^{in,out}\psi^{out} + ((H^\varepsilon - z)^{-1})^{in,in}\psi^{in}$$

and

$$(J(H^{out} - z)^{-1} - (H^\varepsilon - z)^{-1}J)\psi^{out} = \Big(((H^{out} - z)^{-1} - ((H^\varepsilon - z)^{-1})^{out,out})\psi^{out}, -((H^\varepsilon - z)^{-1})^{in,out}\psi^{out}\Big),$$

hence:

By Theorem 1, in the generic case the operator H^ε is ε-quasi unitarily equivalent to the operator $\mathring{H}^{out}$.

By Theorem 2–(iii), in the non-generic case, if the vectors $\underline{\hat{c}}_k$, $k = 1, \ldots, m$, are linearly independent, the operator H^ε is $\varepsilon^{1/2}$-quasi unitarily equivalent to the operator $\widehat{H}^{out}$. More precisely, the second condition in Equation (16) always holds true, while the first one holds true only under the additional assumption that the vectors $\underline{\hat{c}}_k$ are linearly independent.

We refer to [9] for a comprehensive discussion on the comparison between operators acting on different spaces.

3. Kreĭn Resolvent Formulae

In this section we introduce the main tools in our analysis: the Kreĭn-type resolvent formulae for the resolvents of H^ε and $\widehat{H}^{out}$. The proofs are postponed to Appendix A.

Given the Hilbert spaces X^{out}, Y^{out}, X^{in}, and Y^{in}, and a couple of operators $A^{out} : X^{out} \to Y^{out}$ and $A^{in} : X^{in} \to Y^{in}$, we denote by $A := \text{diag}(A^{out}, A^{in})$, the operator $A : X \to Y$, with $X := X^{out} \oplus X^{in}$ and $Y := Y^{out} \oplus Y^{in}$, acting as $Af := (A^{out}f^{out}, A^{in}f^{in})$, for all $f = (f^{out}, f^{in}) \in X$, $f^{out} \in X^{out}$ and $f^{in} \in X^{in}$.

We set

$$D(\mathring{H}^\varepsilon) := D(\mathring{H}^{out}) \oplus D(\mathring{H}^{in,\varepsilon}) \qquad \text{and} \qquad \mathring{H}^\varepsilon := \text{diag}(\mathring{H}^{out}, \mathring{H}^{in,\varepsilon}), \tag{17}$$

with $\mathring{H}^{out}$ and $\mathring{H}^{in,\varepsilon}$ given as in Definitions 6 and 2

Given an operator A, we denote by $\rho(A)$ its resolvent set; the resolvent of A is defined as $(A - z)^{-1}$ for all $z \in \rho(A)$.

For the resolvents of the relevant operators we introduce the shorthand notation

$$R^\varepsilon_z := (H^\varepsilon - z)^{-1} \qquad z \in \rho(H^\varepsilon); \tag{18}$$

$$\mathring{R}^\varepsilon_z := (\mathring{H}^\varepsilon - z)^{-1} \qquad z \in \rho(\mathring{H}^\varepsilon) = \rho(\mathring{H}^{out}) \cap \rho(\mathring{H}^{in,\varepsilon}); \tag{19}$$

$$\mathring{R}_z^{out} := (\mathring{H}^{out} - z)^{-1} \qquad z \in \rho(\mathring{H}^{out}); \qquad \widehat{R}_z^{out} := (\widehat{H}^{out} - z)^{-1} \qquad z \in \rho(\widehat{H}^{out}); \tag{20}$$

$$\mathring{R}_z^{in,\varepsilon} := (\mathring{H}^{in,\varepsilon} - z)^{-1} \qquad z \in \rho(\mathring{H}^{in,\varepsilon}). \tag{21}$$

Obviously, all the operators in Equations (18)–(21) are well-defined and bounded for $z \in \mathbb{C}\backslash\mathbb{R}$, moreover $\mathring{R}_z^\varepsilon = \text{diag}(\mathring{R}_z^{out}, \mathring{R}_z^{in,\varepsilon})$.

Our aim is to write the resolvent difference $R_z^\varepsilon - \mathring{R}_z^\varepsilon$ in a suitable block matrix form, associated to the off-diagonal matrix $\mathbf{\Theta}$ in Equation (29). To do so we follow the approach of Posilicano [22,24]. All the self-adjoint extensions of the symmetric operator obtained by restricting a given self-adjoint operator to the kernel of a given map τ are parametrized by a projection $\mathbf{P}$ and a self-adjoint operator $\mathbf{\Theta}$ in Ran $\mathbf{P}$. We choose the reference operator $\mathring{H}^\varepsilon$ and the map τ so that the Hamiltonian of interest H^ε is the self-adjoint extension parametrized by the identity and the self-adjoint operator given by the off-diagonal matrix $\mathbf{\Theta}$. The Kreĭn formula for the resolvent difference $R_z^\varepsilon - \mathring{R}_z^\varepsilon$, see Lemma 2, is obtained within the approach from [22,24].

We define the maps:

$$\tau^{out} : \mathcal{H}_2^{out} \to \mathbb{C}^N \qquad \tau^{out}\psi := \Psi'(\mathbf{0}); \tag{22}$$

$$\tau^{in} : \mathcal{H}_2^{in,\varepsilon} \to \mathbb{C}^N$$

$$\tau^{in}\psi := \left(\frac{1}{\sqrt{d^{in}(v_1)}} (\mathbf{1}_{d^{in}(v_1)}, \Psi(v_1))_{\mathbb{C}^{d^{in}(v_1)}}, \dots, \frac{1}{\sqrt{d^{in}(v_N)}} (\mathbf{1}_{d^{in}(v_N)}, \Psi(v_N))_{\mathbb{C}^{d^{in}(v_N)}} \right)^T. \tag{23}$$

Moreover we set,

$$\tau : \mathcal{H}_2^\varepsilon = \mathcal{H}_2^{out} \oplus \mathcal{H}_2^{in,\varepsilon} \to \mathbb{C}^{2N} \qquad \tau := \text{diag}(\tau^{out}, \tau^{in}).$$

Note that we are using the identification $\mathbb{C}^{2N} = \mathbb{C}^N \oplus \mathbb{C}^N$.

The following maps are well-defined and bounded

$$\check{G}_z^{out} : \mathcal{H}^{out} \to \mathbb{C}^N \qquad \check{G}_z^{out} := \tau^{out}\mathring{R}_z^{out} \qquad z \in \rho(\mathring{H}^{out})$$

and

$$\check{G}_z^{in,\varepsilon} : \mathcal{H}^{in,\varepsilon} \to \mathbb{C}^N \qquad \check{G}_z^{in,\varepsilon} := \tau^{in}\mathring{R}_z^{in,\varepsilon} \qquad z \in \rho(\mathring{H}^{in,\varepsilon}). \tag{24}$$

Moreover we set

$$\check{G}_z^\varepsilon : \mathcal{H}^\varepsilon = \mathcal{H}^{out} \oplus \mathcal{H}^{in,\varepsilon} \to \mathbb{C}^{2N} \qquad \check{G}_z^\varepsilon := \text{diag}(\check{G}_z^{out}, \check{G}_z^{in,\varepsilon}),$$

for $z \in \rho(\mathring{H}^{out}) \cap \rho(\mathring{H}^{in,\varepsilon})$. Note that $\check{G}_z^\varepsilon = \tau\mathring{R}_z^\varepsilon$ and that all the maps above are well-defined bounded operators for $z \in \mathbb{C}\backslash\mathbb{R}$.

The adjoint maps (in $\bar{z}$) are denoted by

$$G_z^{out} : \mathbb{C}^N \to \mathcal{H}^{out} \qquad G_z^{out} := \check{G}_{\bar{z}}^{out*},$$

$$G_z^{in,\varepsilon} : \mathbb{C}^N \to \mathcal{H}^{in,\varepsilon} \qquad G_z^{in,\varepsilon} := \check{G}_{\bar{z}}^{in,\varepsilon*}, \tag{25}$$

(* denoting the adjoint) and

$$G_z^\varepsilon : \mathbb{C}^{2N} \to \mathcal{H}^\varepsilon \qquad G_z^\varepsilon := \check{G}_{\bar{z}}^{\varepsilon*}.$$

Obviously $G_z^\varepsilon = \text{diag}(G_z^{out}, G_z^{in,\varepsilon})$ to be understood as an operator from $\mathbb{C}^{2N} = \mathbb{C}^N \oplus \mathbb{C}^N$ to $\mathcal{H}^\varepsilon = \mathcal{H}^{out} \oplus \mathcal{H}^{in,\varepsilon}$.

We note that, see Remark A2, $G_z^{out} : \mathbb{C}^N \to \mathcal{H}_2^{out}$ and $G_z^{in,\varepsilon} : \mathbb{C}^N \to \mathcal{H}_2^{in,\varepsilon}$, for all $z \in \rho(\mathring{H}^{out})$ and $z \in \rho(\mathring{H}^{in,\varepsilon})$ respectively, so that the maps ($N \times N$, z-dependent matrices)

$$M_z^{out} : \mathbb{C}^N \to \mathbb{C}^N, \quad M_z^{out} := \tau^{out}G_z^{out} \qquad z \in \rho(\mathring{H}^{out}) \tag{26}$$

$$M_z^{in,\varepsilon} : \mathbb{C}^N \to \mathbb{C}^N, \quad M_z^{in,\varepsilon} := \tau^{in} G_z^{in,\varepsilon} \qquad z \in \rho(\mathring{H}^{in,\varepsilon}), \tag{27}$$

are well defined. Moreover, we set

$$M_z^{\varepsilon} : \mathbb{C}^{2N} \to \mathbb{C}^{2N}, \quad M_z^{\varepsilon} := \operatorname{diag}(M_z^{out}, M_z^{in,\varepsilon}) \qquad z \in \rho(\mathring{H}^{out}) \cap \rho(\mathring{H}^{in,\varepsilon}) = \rho(\mathring{H}^{\varepsilon}); \tag{28}$$

obviously $M_z^{\varepsilon} = \tau G^{\varepsilon}(z)$.

In the following Lemmata we give two Kreĭn-type resolvent formulae: one allows to express the resolvent of $\widehat{H}^{out}$ in terms of the resolvent of $\mathring{H}^{out}$; the other gives the resolvent of H^{ε} in terms of the resolvent of $\mathring{H}^{\varepsilon}$. For the proofs we refer to Appendix A, Appendix A.1.

Lemma 1. *Let $\widehat{P}$ be an orthogonal projection in $\mathbb{C}^N$, and $\widehat{H}^{out}$ and $\mathring{H}^{out}$ be the Hamiltonians defined according to Definitions 7 and 6. Then, for any $z \in \rho(\widehat{H}^{out}) \cap \rho(\mathring{H}^{out})$, the map $\widehat{P} M_z^{out} \widehat{P} : \widehat{P}\mathbb{C}^N \to \widehat{P}\mathbb{C}^N$ is invertible and*

$$\widehat{R}_z^{out} = \mathring{R}_z^{out} - G_z^{out} \widehat{P} \big(\widehat{P} M_z^{out} \widehat{P}\big)^{-1} \widehat{P} \check{G}_z^{out}.$$

Lemma 2. *Let $\boldsymbol{\Theta}$ be the $2N \times 2N$ block matrix*

$$\boldsymbol{\Theta} = \begin{pmatrix} \mathbb{O}_N & \mathbb{I}_N \\ \mathbb{I}_N & \mathbb{O}_N \end{pmatrix}. \tag{29}$$

Then, for any $z \in \rho(H^{\varepsilon}) \cap \rho(\mathring{H}^{\varepsilon})$, the map $(M_z^{\varepsilon} - \boldsymbol{\Theta}) : \mathbb{C}^{2N} \to \mathbb{C}^{2N}$ is invertible and

$$R_z^{\varepsilon} = \mathring{R}_z^{\varepsilon} - G_z^{\varepsilon} \big(M_z^{\varepsilon} - \boldsymbol{\Theta}\big)^{-1} \check{G}_z^{\varepsilon}.$$

We conclude this section with an alternative formula for the resolvent R_z^{ε}. We refer to Appendix A, Appendix A.2, for the proof.

Lemma 3. *Let $z \in \mathbb{C} \backslash \mathbb{R}$, then the maps ($N \times N$, z-dependent matrices)*

$$M_z^{in,\varepsilon} M_z^{out} - \mathbb{I}_N : \mathbb{C}^N \to \mathbb{C}^N \qquad \text{and} \qquad M_z^{out} M_z^{in,\varepsilon} - \mathbb{I}_N : \mathbb{C}^N \to \mathbb{C}^N \tag{30}$$

are invertible. Moreover,

$$R_z^{\varepsilon} = \mathring{R}_z^{\varepsilon} - G_z^{\varepsilon} \begin{pmatrix} \big(M_z^{in,\varepsilon} M_z^{out} - \mathbb{I}_N\big)^{-1} M_z^{in,\varepsilon} & \big(M_z^{in,\varepsilon} M_z^{out} - \mathbb{I}_N\big)^{-1} \\ \big(M_z^{out} M_z^{in,\varepsilon} - \mathbb{I}_N\big)^{-1} & M_z^{out} \big(M_z^{in,\varepsilon} M_z^{out} - \mathbb{I}_N\big)^{-1} \end{pmatrix} \check{G}_z^{\varepsilon}. \tag{31}$$

4. Scale Invariance

In this section we discuss the scale invariance properties of $\mathring{H}^{in,\varepsilon}$ and collect several formulae concerning the operators $\mathring{R}_z^{in,\varepsilon}$, $\check{G}_z^{in,\varepsilon}$, $G_z^{in,\varepsilon}$, and $M_z^{in,\varepsilon}$.

Recall that we have denoted by λ_n and $\{\varphi_n\}_{n\in\mathbb{N}}$ the eigenvalues and a corresponding set of orthonormal eigenfunctions of $\mathring{H}^{in}$.

The eigenvalues of $\mathring{H}^{in,\varepsilon}$ (counting multiplicity) and a corresponding set of orthonormal eigenfunctions are given by

$$\lambda_n^{\varepsilon} = \varepsilon^{-2} \lambda_n\,; \qquad \varphi_n^{\varepsilon}(x) = \varepsilon^{-1/2} \varphi_n(x/\varepsilon), \tag{32}$$

where λ_n are the eigenvalues of $\mathring{H}^{in}$, and φ_n the corresponding (orthonormal) eigenfunctions.

By the spectral theorem and by the scaling properties (32), $\mathring{R}_z^{in,\varepsilon}$ is given by

$$\mathring{R}_z^{in,\varepsilon} = \sum_{n\in\mathbb{N}} \frac{\varphi_n^{\varepsilon} (\varphi_n^{\varepsilon}, \cdot)_{\mathcal{H}^{in,\varepsilon}}}{\lambda_n^{\varepsilon} - z} = \varepsilon^2 \sum_{n\in\mathbb{N}} \frac{\varphi_n^{\varepsilon} (\varphi_n^{\varepsilon}, \cdot)_{\mathcal{H}^{in,\varepsilon}}}{\lambda_n - \varepsilon^2 z}. \tag{33}$$

Hence, its integral kernel can be written as

$$\mathring{R}_z^{in,\varepsilon}(x,y) = \varepsilon \sum_{n\in\mathbb{N}} \frac{\varphi_n(x/\varepsilon)\varphi_n(y/\varepsilon)}{\lambda_n - \varepsilon^2 z} \qquad x,y \in \mathcal{G}^{in,\varepsilon}. \tag{34}$$

Since there exists a positive constant C such that $\sup_{x\in\mathcal{G}^{in}} |\varphi_n(x)| \leq C$ and $\lambda_n \geq Cn^2$ for n large enough (see Appendix B), the series in Equation (34) is uniformly convergent for $x, y \in \mathcal{G}^{in,\varepsilon}$. Hence, we can write the operators $\check{G}_z^{in,\varepsilon}$ and $G_z^{in,\varepsilon}$, and the matrix $M_z^{in,\varepsilon}$ in a similar way, see Equations (35) and (36) below.

Note that, since functions in $D(\mathring{H}^{in,\varepsilon})$ are continuous in the connecting vertices, the eigenfunctions φ_n^ε can be evaluated in the connecting vertices, and, by the definition of τ^{in} (see Equation (23)), one has

$$\tau^{in}\varphi_n^\varepsilon = (\varphi_n^\varepsilon(v_1), \ldots, \varphi_n^\varepsilon(v_N))^T.$$

So that, for any eigenfunction φ_n^ε we can define the vector $\underline{c}_n^\varepsilon$ as

$$\underline{c}_n^\varepsilon := \tau^{in}\varphi_n^\varepsilon.$$

We note that $\underline{c}_n^\varepsilon = \varepsilon^{-1/2}\underline{c}_n$, with

$$\underline{c}_n = (\varphi_n(v_1), \ldots, \varphi_n(v_N))^T,$$

and that the vectors $\underline{c}_n$ are defined in the same way as the vectors $\hat{\underline{c}}_k$ in Equation (9).

Remark 7. *In the non-generic case, zero is an eigenvalue of $\mathring{H}^{in,\varepsilon}$. We denote by $\{\hat{\varphi}_k^\varepsilon\}_{k=1,\ldots,m}$ the corresponding set of (orthonormal) eigenfunctions given by $\hat{\varphi}_k^\varepsilon(x) = \varepsilon^{-1/2}\hat{\varphi}_k(x/\varepsilon)$ where $\hat{\varphi}_k$ are the eigenfunctions corresponding to the eigenvalue zero of $\mathring{H}^{in}$. The vectors $\hat{\underline{c}}_k^\varepsilon := \tau^{in}\hat{\varphi}_k^\varepsilon$ are related to the vectors $\hat{\underline{c}}_k$ by the identity $\hat{\underline{c}}_k^\varepsilon = \varepsilon^{-1/2}\hat{\underline{c}}_k$.*

By the discussion above, and by the definitions (24), (25), and (27), we obtain

$$\check{G}_z^{in,\varepsilon} = \varepsilon^{3/2} \sum_{n\in\mathbb{N}} \frac{\underline{c}_n(\varphi_n^\varepsilon,\cdot)_{\mathcal{H}^{in,\varepsilon}}}{\lambda_n - \varepsilon^2 z}; \qquad G_z^{in,\varepsilon} = \varepsilon^{3/2} \sum_{n\in\mathbb{N}} \frac{\varphi_n^\varepsilon(\underline{c}_n,\cdot)_{\mathbb{C}^N}}{\lambda_n - \varepsilon^2 z}, \tag{35}$$

and

$$M_z^{in,\varepsilon} = \varepsilon \sum_{n\in\mathbb{N}} \frac{\underline{c}_n(\underline{c}_n,\cdot)_{\mathbb{C}^N}}{\lambda_n - \varepsilon^2 z}. \tag{36}$$

5. Proof of Theorems 1 and 2

This section is devoted to the proofs of Theorems 1 and 2. Actually, we shall prove a finer version of the results with more precise estimates of the remainders, see Theorems 3 and 4 below.

Remark 8. *By Equation (31), it follows that, in the out/in decomposition (11), the resolvent R_z^ε can be written as*

$$R_z^\varepsilon = \begin{pmatrix} \mathring{R}_z^{out} & \mathbb{O} \\ \mathbb{O} & \mathring{R}_z^{in,\varepsilon} \end{pmatrix} - \begin{pmatrix} \mathcal{R}_z^{out,out,\varepsilon} & \mathcal{R}_z^{out,in,\varepsilon} \\ \mathcal{R}_z^{in,out,\varepsilon} & \mathcal{R}_z^{in,in,\varepsilon} \end{pmatrix} \tag{37}$$

with

$$\mathcal{R}_z^{out,out,\varepsilon} = G_z^{out}\left(M_z^{in,\varepsilon}M_z^{out} - \mathbb{I}_N\right)^{-1} M_z^{in,\varepsilon}\check{G}_z^{out}; \tag{38}$$

$$\mathcal{R}_z^{in,out,\varepsilon} = G_z^{in,\varepsilon}\left(M_z^{out}M_z^{in,\varepsilon} - \mathbb{I}_N\right)^{-1}\check{G}_z^{out}; \tag{39}$$

$$\mathcal{R}_z^{out,in,\varepsilon} = G_z^{out}\left(M_z^{in,\varepsilon}M_z^{out} - \mathbb{I}_N\right)^{-1}\check{G}_z^{in,\varepsilon}; \tag{40}$$

$$\mathcal{R}_z^{in,in,\varepsilon} = G_z^{in,\varepsilon}M_z^{out}\left(M_z^{in,\varepsilon}M_z^{out} - \mathbb{I}_N\right)^{-1}\check{G}_z^{in,\varepsilon}. \tag{41}$$

Note that since $M_z = M_{\bar{z}}^*$ holds true both for the "out" and "in" M-matrices (see Equation (A2)), one infers $\mathcal{R}_z^{in,out,\varepsilon} = \mathcal{R}_{\bar{z}}^{out,in,\varepsilon*}$.

5.1. Generic Case. Proof of Theorem 1

In this section we study the limit of the relevant quantities in the generic case and prove Theorem 1.

Proposition 1. *Let $z \in \mathbb{C}\backslash\mathbb{R}$. In the generic case,*

$$\mathring{R}_z^{in,\varepsilon} = \mathcal{O}_{\mathcal{B}(\mathcal{H}^{in,\varepsilon})}(\varepsilon^2); \tag{42}$$

$$\check{G}_z^{in,\varepsilon} = \mathcal{O}_{\mathcal{B}(\mathcal{H}^{in,\varepsilon},\mathbb{C}^N)}(\varepsilon^{3/2}) ; \qquad G_z^{in,\varepsilon} = \mathcal{O}_{\mathcal{B}(\mathbb{C}^N,\mathcal{H}^{in,\varepsilon})}(\varepsilon^{3/2}). \tag{43}$$

Proof. We prove first Claim (42). For any $\psi^{in} \in \mathcal{H}^{in,\varepsilon}$, since $\{\varphi_n^\varepsilon\}_{n\in\mathbb{N}}$ is an orthonormal set of eigenfunctions in $\mathcal{H}^{in,\varepsilon}$, and by Equation (33), we infer

$$\|\mathring{R}_z^{in,\varepsilon}\psi^{in}\|_{\mathcal{H}^{in,\varepsilon}} = \varepsilon^2 \left(\sum_{n\in\mathbb{N}} \frac{|(\varphi_n^\varepsilon, \psi^{in})_{\mathcal{H}^{in,\varepsilon}}|^2}{|\lambda_n - \varepsilon^2 z|^2} \right)^{1/2} \leq C\varepsilon^2 \|\psi^{in}\|_{\mathcal{H}^{in,\varepsilon}},$$

where in the latter inequality we used the bound $|\lambda_n - \varepsilon^2 z|^{-2} \leq 4|\lambda_n|^{-2} \leq C$, which holds true in the generic case because $|\lambda_n - \varepsilon^2 z| \geq |\lambda_n|/2 \geq C$ for all $n \in \mathbb{N}$ and ε small enough.

To prove the first claim in Equation (43), let $\psi^{in} \in \mathcal{H}^{in,\varepsilon}$, then

$$\check{G}_z^{in,\varepsilon}\psi^{in} = \varepsilon^{3/2} \sum_{n\in\mathbb{N}} \frac{\underline{c}_n (\varphi_n^\varepsilon, \psi^{in})_{\mathcal{H}^{in,\varepsilon}}}{\lambda_n - \varepsilon^2 z}.$$

Hence, from the Cauchy–Schwarz inequality,

$$\begin{aligned}\|\check{G}_z^{in,\varepsilon}\psi^{in}\|_{\mathbb{C}^N} &\leq \varepsilon^{3/2} \sum_{n\in\mathbb{N}} \frac{\|\underline{c}_n\|_{\mathbb{C}^N}\, |(\varphi_n^\varepsilon, \psi^{in})_{\mathcal{H}^{in,\varepsilon}}|}{|\lambda_n - \varepsilon^2 z|} \\ &\leq \varepsilon^{3/2} \|\psi^{in}\|_{\mathcal{H}^{in,\varepsilon}} \left(\sum_{n\in\mathbb{N}} \frac{\|\underline{c}_n\|^2_{\mathbb{C}^N}}{|\lambda_n - \varepsilon^2 z|^2} \right)^{1/2} \leq C\,\varepsilon^{3/2} \|\psi^{in}\|_{\mathcal{H}^{in,\varepsilon}},\end{aligned}$$

because $\|\underline{c}_n\|^2_{\mathbb{C}^N} \leq C$ and $\sum_{n\in\mathbb{N}} |\lambda_n - \varepsilon^2 z|^{-2} \leq C \sum_{n\in\mathbb{N}} |\lambda_n|^{-2} \leq C$. This proves the first Claim in Equation (43); the second one is trivial, being $G_z^{in,\varepsilon}$ the adjoint of $\check{G}_{\bar{z}}^{in,\varepsilon}$. □

Proposition 2. *Let $z \in \mathbb{C}\backslash\mathbb{R}$. In the generic case,*

$$M_z^{in,\varepsilon} = \mathcal{O}_{\mathcal{B}(\mathbb{C}^N)}(\varepsilon). \tag{44}$$

Proof. Recall Equation (36) and note that for any $\underline{q} \in \mathbb{C}^N$,

$$\|M_z^{in,\varepsilon}\underline{q}\|_{\mathbb{C}^N} \leq \varepsilon \sum_{n\in\mathbb{N}} \frac{\|\underline{c}_n\|_{\mathbb{C}^N}\, |(\underline{c}_n, \underline{q})_{\mathbb{C}^N}|}{|\lambda_n - \varepsilon^2 z|} \leq \varepsilon \|\underline{q}\|_{\mathbb{C}^N} \sum_{n\in\mathbb{N}} \frac{\|\underline{c}_n\|^2_{\mathbb{C}^N}}{|\lambda_n - \varepsilon^2 z|} \leq C\varepsilon \|\underline{q}\|_{\mathbb{C}^N},$$

because $\|\underline{c}_n\|^2_{\mathbb{C}^N} \leq C$ and $\sum_{n\in\mathbb{N}} |\lambda_n - \varepsilon^2 z|^{-1} \leq C \sum_{n\in\mathbb{N}} |\lambda_n|^{-1} \leq C$. □

Theorem 3. *Let $z \in \mathbb{C}\backslash\mathbb{R}$. In the generic case*

$$R_z^\varepsilon = \begin{pmatrix} \mathring{R}_z^{out} + \mathcal{O}_{\mathcal{B}(\mathcal{H}^{out})}(\varepsilon) & \mathcal{O}_{\mathcal{B}(\mathcal{H}^{in,\varepsilon},\mathcal{H}^{out})}(\varepsilon^{3/2}) \\ \mathcal{O}_{\mathcal{B}(\mathcal{H}^{out},\mathcal{H}^{in,\varepsilon})}(\varepsilon^{3/2}) & \mathcal{O}_{\mathcal{B}(\mathcal{H}^{in,\varepsilon})}(\varepsilon^2) \end{pmatrix}, \tag{45}$$

where the expansion has to be understood in the out / in decomposition (11).

Proof. Note that $\left(M_z^{in,\varepsilon} M_z^{out} - \mathbb{I}_N\right)^{-1} = \mathcal{O}_{\mathbb{C}^N}(1)$ by Equation (44) and because M_z^{out} is bounded and does not depend on ε. Hence, $\left(M_z^{in,\varepsilon} M_z^{out} - \mathbb{I}_N\right)^{-1} M_z^{in,\varepsilon} = \mathcal{O}_{\mathbb{C}^N}(\varepsilon)$.

To conclude, by Equations (38)–(41), and by expansions (43), we infer: $\mathcal{R}_z^{out,out,\varepsilon} = \mathcal{O}_{\mathcal{B}(\mathcal{H}^{out})}(\varepsilon)$; $\mathcal{R}_z^{out,in,\varepsilon} = \mathcal{O}_{\mathcal{B}(\mathcal{H}^{in,\varepsilon},\mathcal{H}^{out})}(\varepsilon^{3/2})$; $\mathcal{R}_z^{in,out,\varepsilon} = \mathcal{O}_{\mathcal{B}(\mathcal{H}^{out},\mathcal{H}^{in,\varepsilon})}(\varepsilon^{3/2})$ (this is obvious since it is the adjoint of $\mathcal{R}_{\bar{z}}^{out,in,\varepsilon}$); and $\mathcal{R}_z^{in,in,\varepsilon} = \mathcal{O}_{\mathcal{B}(\mathcal{H}^{in,\varepsilon})}(\varepsilon^3)$.

Expansion (45) follows by taking into account the bound (42), and from Remark 8. □

Theorem 1 is a direct consequence of Theorem 3.

5.2. Non-Generic Case. Proof of Theorem 2

In this section we study the limit of the relevant quantities in the non-generic case and prove Theorem 2.

Recall that, in the non-generic case, $\{\hat{\varphi}_k^\varepsilon\}_{k=1,\ldots,m}$ denotes a set of orthonormal eigenfunctions corresponding to the zero eigenvalue, see also Remark 7.

Proposition 3. *Let $z \in \mathbb{C}\backslash\mathbb{R}$. In the non-generic case*

$$\mathring{R}_z^{in,\varepsilon} = -\sum_{k=1}^{m} \frac{\hat{\varphi}_k^\varepsilon(\hat{\varphi}_{k'}^\varepsilon,\cdot)_{\mathcal{H}^{in,\varepsilon}}}{z} + \mathcal{O}_{\mathcal{B}(\mathcal{H}^{in,\varepsilon})}(\varepsilon^2); \tag{46}$$

$$\check{G}_z^{in,\varepsilon} = -\sum_{k=1}^{m} \frac{\hat{c}_k(\hat{\varphi}_{k'}^\varepsilon,\cdot)_{\mathcal{H}^{in,\varepsilon}}}{\varepsilon^{1/2}z} + \mathcal{O}_{\mathcal{B}(\mathcal{H}^{in,\varepsilon},\mathbb{C}^N)}(\varepsilon^{3/2}); \tag{47}$$

$$G_z^{in,\varepsilon} = -\sum_{k=1}^{m} \frac{\hat{\varphi}_k^\varepsilon(\hat{c}_{k'},\cdot)_{\mathbb{C}^N}}{\varepsilon^{1/2}z} + \mathcal{O}_{\mathcal{B}(\mathbb{C}^N,\mathcal{H}^{in,\varepsilon})}(\varepsilon^{3/2}). \tag{48}$$

Proof. We prove first Claim (46). By Equation (33) we infer

$$\mathring{R}_z^{in,\varepsilon} = -\sum_{k=1}^{m} \frac{\hat{\varphi}_k^\varepsilon(\hat{\varphi}_{k'}^\varepsilon,\cdot)_{\mathcal{H}^{in,\varepsilon}}}{z} + \varepsilon^2 \sum_{n:\lambda_n\neq 0} \frac{\varphi_n^\varepsilon(\varphi_{n'}^\varepsilon,\cdot)_{\mathcal{H}^{in,\varepsilon}}}{\lambda_n - \varepsilon^2 z}. \tag{49}$$

Note that the second sum runs over $\lambda_n \neq 0$, hence one has the bound $|\lambda_n - \varepsilon^2 z| \geq |\lambda_n|/2 \geq C$, for ε small enough. For this reason, the bound in Equation (46) on the second term at the r.h.s. of Equation (49) can be obtained with an argument similar to the one used in the proof of bound (42).

To prove Claim (47) we proceed in a similar way. We note that, see Equation (35),

$$\check{G}_z^{in,\varepsilon} = -\sum_{k=1}^{m} \frac{\hat{c}_k(\hat{\varphi}_{k'}^\varepsilon,\cdot)_{\mathcal{H}^{in,\varepsilon}}}{\varepsilon^{1/2}z} + \varepsilon^{3/2} \sum_{n:\lambda_n\neq 0} \frac{c_n(\varphi_{n'}^\varepsilon,\cdot)_{\mathcal{H}^{in,\varepsilon}}}{\lambda_n - \varepsilon^2 z},$$

and bound the second term at the r.h.s. by reasoning in the same way as in the proof of Proposition 1. Claim (48) follows by noticing that $G_z^{in,\varepsilon}$ is the adjoint of $\check{G}_{\bar{z}}^{in,\varepsilon}$. □

Next we prove a proposition on the expansion of the $N \times N$, z-dependent matrix $M_z^{in,\varepsilon}$. Recall that $\widehat{C}$ was defined in Definition 5.

Proposition 4. *Let $z \in \mathbb{C}\backslash\mathbb{R}$. In the non-generic case,*

$$M_z^{in,\varepsilon} = -\frac{1}{\varepsilon z}\widehat{C} + \mathcal{O}_{\mathcal{B}(\mathbb{C}^N)}(\varepsilon). \tag{50}$$

Proof. The claim immediately follows from Equation (36), after noticing that

$$M_z^{in,\varepsilon} = -\frac{1}{\varepsilon z}\widehat{C} + \varepsilon \sum_{n:\lambda_n \neq 0} \frac{\underline{c}_n(\underline{c}_n, \cdot)_{\mathbb{C}^N}}{\lambda_n - \varepsilon^2 z}$$

and by treating the second term at the r.h.s. with argument similar to the one used in the proof of Proposition 2. □

We set

$$\widetilde{M}_z^{in,\varepsilon} := \varepsilon M_z^{in,\varepsilon}$$

and recall that M_z^{out} is invertible (see Remark A3), then

$$(M_z^{in,\varepsilon} M_z^{out} - \mathbb{I}_N)^{-1} = \varepsilon M_z^{out^{-1}} (\widetilde{M}_z^{in,\varepsilon} - \varepsilon M_z^{out^{-1}})^{-1}. \tag{51}$$

In the following proposition we give an expansion formula for the term $(\widetilde{M}_z^{in,\varepsilon} - \varepsilon M_z^{out^{-1}})^{-1}$ in the non-generic case.

Proposition 5. *Let $z \in \mathbb{C}\backslash\mathbb{R}$. In the non-generic case, decompose the space $\mathbb{C}^N$ as $\mathbb{C}^N = \widehat{P}\mathbb{C}^N \oplus \widehat{P}^\perp\mathbb{C}^N$, and denote by $\widehat{C}_0$ the restriction of $\widehat{C}$ to $\widehat{P}\mathbb{C}^N$. Then, the map $\widehat{P}^\perp M_z^{out^{-1}} \widehat{P}^\perp$ is invertible in $\widehat{P}^\perp\mathbb{C}^N$.*

Set

$$N_z := (\widehat{P}^\perp M_z^{out^{-1}} \widehat{P}^\perp)^{-1} : \widehat{P}^\perp\mathbb{C}^N \to \widehat{P}^\perp\mathbb{C}^N, \tag{52}$$

then

$$\begin{aligned}&(\widetilde{M}_z^{in,\varepsilon} - \varepsilon M_z^{out^{-1}})^{-1} \\ &= -\begin{pmatrix} z\widehat{C}_0^{-1} + \mathcal{O}_{\mathcal{B}(\widehat{P}\mathbb{C}^N)}(\varepsilon) & -z\widehat{C}_0^{-1}\widehat{P}M_z^{out^{-1}}\widehat{P}^\perp N_z + \mathcal{O}_{\mathcal{B}(\widehat{P}^\perp\mathbb{C}^N,\widehat{P}\mathbb{C}^N)}(\varepsilon) \\ -zN_z\widehat{P}^\perp M_z^{out^{-1}}\widehat{P}\widehat{C}_0^{-1} + \mathcal{O}_{\mathcal{B}(\widehat{P}\mathbb{C}^N,\widehat{P}^\perp\mathbb{C}^N)}(\varepsilon) & \varepsilon^{-1}N_z + \mathcal{O}_{\mathcal{B}(\widehat{P}^\perp\mathbb{C}^N)}(1)\end{pmatrix},\end{aligned} \tag{53}$$

to be understood in the decomposition $\mathbb{C}^N = \widehat{P}\mathbb{C}^N \oplus \widehat{P}^\perp\mathbb{C}^N$.

Proof. We postpone the proof of the fact that the map $\widehat{P}^\perp M_z^{out^{-1}} \widehat{P}^\perp$ is invertible in $\widehat{P}^\perp\mathbb{C}^N$ to the appendix, see Remark A4.

Next we prove that the expansion formula (53) holds true. We start by noticing that the map $z^{-1}\widehat{C} + \varepsilon M_z^{out^{-1}}$ is invertible. In fact, by Remark A4 and since $(\underline{q}, \widehat{C}\underline{q})_{\mathbb{C}^N} = \sum_{k=1}^m |(\underline{\hat{c}}_k, \underline{q})_{\mathbb{C}^N}|^2 \geq 0$, we infer

$$\operatorname{Im}\left(\underline{q}, (z^{-1}\widehat{C} + \varepsilon M_z^{out^{-1}})\underline{q}\right)_{\mathbb{C}^N} = -\frac{\operatorname{Im} z}{|z|^2}(\underline{q}, \widehat{C}\underline{q})_{\mathbb{C}^N} - \varepsilon \operatorname{Im} z \|G_z^{out} M_z^{out^{-1}} \underline{q}\|^2_{\mathcal{H}^{out}} \neq 0,$$

because it is the sum of two non-positive (or non-negative) terms and $\|G_z^{out} M_z^{out^{-1}} \underline{q}\|^2_{\mathcal{H}^{out}} \neq 0$ by the injectivity of $G_z^{out} M_z^{out^{-1}}$, see Remark A1.

Moreover we have the a-priori estimate

$$(\widetilde{M}_z^{in,\varepsilon} - \varepsilon M_z^{out^{-1}})^{-1} = \mathcal{O}_{\mathcal{B}(\mathbb{C}^N)}(\varepsilon^{-1}). \tag{54}$$

The latter follows from (see also Equation (A3))

$$\begin{aligned}\|\underline{q}\|_{\mathbb{C}^N} \|(\widetilde{M}_z^{in,\varepsilon} - \varepsilon M_z^{out^{-1}})\underline{q}\|_{\mathbb{C}^N} \geq & |(\underline{q}, \widetilde{M}_z^{in,\varepsilon} - \varepsilon M_z^{out^{-1}}\underline{q})_{\mathbb{C}^N}| \\ \geq & |\operatorname{Im}(\underline{q}, \widetilde{M}_z^{in,\varepsilon} - \varepsilon M_z^{out^{-1}}\underline{q})_{\mathbb{C}^N}| \\ = & \varepsilon|\operatorname{Im} z|(\|G_z^{in,\varepsilon}\underline{q}\|^2_{\mathcal{H}^{in,\varepsilon}} + \|G_z^{out} M_z^{out^{-1}}\underline{q}\|^2_{\mathcal{H}^{out}}) \geq \varepsilon C_z \|\underline{q}\|^2_{\mathbb{C}^N},\end{aligned}$$

for some positive constant C_z, from the injectivity of $G_z^{out}M_z^{out^{-1}}$. Hence, setting $\underline{q} = (\widetilde{M}_z^{in,\varepsilon} - \varepsilon M_z^{out^{-1}})^{-1}\underline{p}$, it follows that $\|(\widetilde{M}_z^{in,\varepsilon} - \varepsilon M_z^{out^{-1}})^{-1}\underline{p}\|_{\mathbb{C}^N} \leq (\varepsilon C_z)^{-1}\|\underline{p}\|_{\mathbb{C}^N}$.

Next we use the expansion (see Equation (50))

$$\widetilde{M}_z^{in,\varepsilon} = -\frac{1}{z}\widehat{C} + \mathcal{O}_{\mathcal{B}(\mathbb{C}^N)}(\varepsilon^2), \tag{55}$$

which, together with the a-priori estimate (54), gives

$$\begin{aligned}(\widetilde{M}_z^{in,\varepsilon} - \varepsilon M_z^{out^{-1}})^{-1} &= -(z^{-1}\widehat{C} + \varepsilon M_z^{out^{-1}})^{-1} + (z^{-1}\widehat{C} + \varepsilon M_z^{out^{-1}})^{-1}\mathcal{O}_{\mathcal{B}(\mathbb{C}^N)}(\varepsilon^2)(\widetilde{M}_z^{in,\varepsilon} - \varepsilon M_z^{out^{-1}})^{-1} \\ &= -(z^{-1}\widehat{C} + \varepsilon M_z^{out^{-1}})^{-1} + (z^{-1}\widehat{C} + \varepsilon M_z^{out^{-1}})^{-1}\mathcal{O}_{\mathcal{B}(\mathbb{C}^N)}(\varepsilon).\end{aligned} \tag{56}$$

Here we used the formula $(A+B)^{-1} = A^{-1} - A^{-1}B(A+B)^{-1}$. Note that by using instead the complementary formula $(A+B)^{-1} = A^{-1} - (A+B)^{-1}BA^{-1}$, we obtain

$$(\widetilde{M}_z^{in,\varepsilon} - \varepsilon M_z^{out^{-1}})^{-1} = -(z^{-1}\widehat{C} + \varepsilon M_z^{out^{-1}})^{-1} + \mathcal{O}_{\mathcal{B}(\mathbb{C}^N)}(\varepsilon)(z^{-1}\widehat{C} + \varepsilon M_z^{out^{-1}})^{-1}. \tag{57}$$

Next we analyze the term $(z^{-1}\widehat{C} + \varepsilon M_z^{out^{-1}})^{-1}$.

We start by noticing that the map $z^{-1}\widehat{C}_0 + \varepsilon\widehat{P}M_z^{out^{-1}}\widehat{P} : \widehat{P}\mathbb{C}^N \to \widehat{P}\mathbb{C}^N$ is invertible, because $\widehat{C}_0$ is invertible in $\widehat{P}\mathbb{C}^N$ and $\varepsilon\widehat{P}M_z^{out^{-1}}\widehat{P} = \mathcal{O}_{\mathbb{C}^N}(\varepsilon)$.

By the identification (to be understood in the decomposition $\mathbb{C}^N = \widehat{P}\mathbb{C}^N \oplus \widehat{P}^\perp\mathbb{C}^N$)

$$M_z^{out^{-1}} = \begin{pmatrix} \widehat{P}M_z^{out^{-1}}\widehat{P} & \widehat{P}M_z^{out^{-1}}\widehat{P}^\perp \\ \widehat{P}^\perp M_z^{out^{-1}}\widehat{P} & \widehat{P}^\perp M_z^{out^{-1}}\widehat{P}^\perp \end{pmatrix}, \tag{58}$$

we have the identity

$$z^{-1}\widehat{C} + \varepsilon M_z^{out^{-1}} = \begin{pmatrix} z^{-1}\widehat{C}_0 + \varepsilon\widehat{P}M_z^{out^{-1}}\widehat{P} & \varepsilon\widehat{P}M_z^{out^{-1}}\widehat{P}^\perp \\ \varepsilon\widehat{P}^\perp M_z^{out^{-1}}\widehat{P} & \varepsilon\widehat{P}^\perp M_z^{out^{-1}}\widehat{P}^\perp \end{pmatrix}.$$

Hence, from the block-matrix inversion formula, we obtain

$$(z^{-1}\widehat{C} + \varepsilon M_z^{out^{-1}})^{-1} = \begin{pmatrix} D_z^\varepsilon & -D_z^\varepsilon\widehat{P}M_z^{out^{-1}}\widehat{P}^\perp N_z \\ -N_z\widehat{P}^\perp M_z^{out^{-1}}\widehat{P}D_z^\varepsilon & \varepsilon^{-1}N_z + N_z\widehat{P}^\perp M_z^{out^{-1}}\widehat{P}D_z^\varepsilon\widehat{P}M_z^{out^{-1}}\widehat{P}^\perp N_z \end{pmatrix},$$

with $D_z^\varepsilon : \widehat{P}\mathbb{C}^N \to \widehat{P}\mathbb{C}^N$ given by

$$D_z^\varepsilon := \left(z^{-1}\widehat{C}_0 + \varepsilon\widehat{P}M_z^{out^{-1}}\widehat{P} - \varepsilon\widehat{P}M_z^{out^{-1}}\widehat{P}^\perp(\widehat{P}^\perp M_z^{out^{-1}}\widehat{P}^\perp)^{-1}\widehat{P}^\perp M_z^{out^{-1}}\widehat{P}\right)^{-1};$$

note that D_z^ε is well-defined because it is the inverse of a map of the form $z^{-1}\widehat{C}_0 + \mathcal{O}_{\mathcal{B}(\widehat{P}\mathbb{C}^N)}(\varepsilon)$, and $z^{-1}\widehat{C}_0$ is invertible in $\widehat{P}\mathbb{C}^N$.

Moreover, it holds true,

$$D_z^\varepsilon = z\widehat{C}_0^{-1} + \mathcal{O}_{\mathcal{B}(\widehat{P}\mathbb{C}^N)}(\varepsilon).$$

Hence,

$$\begin{aligned}&(z^{-1}\widehat{C} + \varepsilon M_z^{out^{-1}})^{-1} \\ &= \begin{pmatrix} z\widehat{C}_0^{-1} & -z\widehat{C}_0^{-1}\widehat{P}M_z^{out^{-1}}\widehat{P}^\perp N_z \\ -zN_z\widehat{P}^\perp M_z^{out^{-1}}\widehat{P}\widehat{C}_0^{-1} & \varepsilon^{-1}N_z + zN_z\widehat{P}^\perp M_z^{out^{-1}}\widehat{P}\widehat{C}_0^{-1}\widehat{P}M_z^{out^{-1}}\widehat{P}^\perp N_z \end{pmatrix} + \mathcal{O}_{\mathcal{B}(\mathbb{C}^N)}(\varepsilon).\end{aligned}$$

The latter can also be written as

$$(z^{-1}\widehat{C} + \varepsilon M_z^{out^{-1}})^{-1} = \begin{pmatrix} z\widehat{C}_0^{-1} & -z\widehat{C}_0^{-1}\widehat{P}M_z^{out^{-1}}\widehat{P}^{\perp}N_z \\ -zN_z\widehat{P}^{\perp}M_z^{out^{-1}}\widehat{P}\widehat{C}_0^{-1} & \varepsilon^{-1}N_z + \mathcal{O}_{\mathcal{B}(\widehat{P}^{\perp}\mathbb{C}^N)}(1) \end{pmatrix} + \mathcal{O}_{\mathcal{B}(\mathbb{C}^N)}(\varepsilon).$$

Using this expansion formula in Equation (56) we obtain

$$\begin{aligned}
&(\widetilde{M}_z^{in,\varepsilon} - \varepsilon M_z^{out^{-1}})^{-1} \\
&= -\begin{pmatrix} z\widehat{C}_0^{-1} & -z\widehat{C}_0^{-1}\widehat{P}M_z^{out^{-1}}\widehat{P}^{\perp}N_z \\ -zN_z\widehat{P}^{\perp}M_z^{out^{-1}}\widehat{P}\widehat{C}_0^{-1} & \varepsilon^{-1}N_z + \mathcal{O}_{\mathcal{B}(\widehat{P}^{\perp}\mathbb{C}^N)}(1) \end{pmatrix} \\
&\quad + \begin{pmatrix} z\widehat{C}_0^{-1} & -z\widehat{C}_0^{-1}\widehat{P}M_z^{out^{-1}}\widehat{P}^{\perp}N_z \\ -zN_z\widehat{P}^{\perp}M_z^{out^{-1}}\widehat{P}\widehat{C}_0^{-1} & \varepsilon^{-1}N_z + \mathcal{O}_{\mathcal{B}(\widehat{P}^{\perp}\mathbb{C}^N)}(1) \end{pmatrix} \mathcal{O}_{\mathcal{B}(\mathbb{C}^N)}(\varepsilon) + \mathcal{O}_{\mathcal{B}(\mathbb{C}^N)}(\varepsilon) \\
&= -\begin{pmatrix} z\widehat{C}_0^{-1} + \mathcal{O}_{\mathcal{B}(\widehat{P}\mathbb{C}^N)}(\varepsilon) & -z\widehat{C}_0^{-1}\widehat{P}M_z^{out^{-1}}\widehat{P}^{\perp}N_z + \mathcal{O}_{\mathcal{B}(\widehat{P}^{\perp}\mathbb{C}^N,\widehat{P}\mathbb{C}^N)}(\varepsilon) \\ \mathcal{O}_{\mathcal{B}(\widehat{P}\mathbb{C}^N,\widehat{P}^{\perp}\mathbb{C}^N)}(1) & \varepsilon^{-1}N_z + \mathcal{O}_{\mathcal{B}(\widehat{P}^{\perp}\mathbb{C}^N)}(1) \end{pmatrix}.
\end{aligned}$$

On the other hand, using Equation (57), we obtain

$$\begin{aligned}
(\widetilde{M}_z^{in,\varepsilon} - \varepsilon M_z^{out^{-1}})^{-1} &= -\begin{pmatrix} z\widehat{C}_0^{-1} & -z\widehat{C}_0^{-1}\widehat{P}M_z^{out^{-1}}\widehat{P}^{\perp}N_z \\ -zN_z\widehat{P}^{\perp}M_z^{out^{-1}}\widehat{P}\widehat{C}_0^{-1} & \varepsilon^{-1}N_z + \mathcal{O}_{\mathcal{B}(\widehat{P}^{\perp}\mathbb{C}^N)}(1) \end{pmatrix} \\
&\quad + \mathcal{O}_{\mathcal{B}(\mathbb{C}^N)}(\varepsilon)\begin{pmatrix} z\widehat{C}_0^{-1} & -z\widehat{C}_0^{-1}\widehat{P}M_z^{out^{-1}}\widehat{P}^{\perp}N_z \\ -zN_z\widehat{P}^{\perp}M_z^{out^{-1}}\widehat{P}\widehat{C}_0^{-1} & \varepsilon^{-1}N_z + \mathcal{O}_{\mathcal{B}(\widehat{P}^{\perp}\mathbb{C}^N)}(1) \end{pmatrix} + \mathcal{O}_{\mathcal{B}(\mathbb{C}^N)}(\varepsilon) \\
&= -\begin{pmatrix} z\widehat{C}_0^{-1} + \mathcal{O}_{\mathcal{B}(\widehat{P}\mathbb{C}^N)}(\varepsilon) & \mathcal{O}_{\mathcal{B}(\widehat{P}^{\perp}\mathbb{C}^N,\widehat{P}\mathbb{C}^N)}(1) \\ -zN_z\widehat{P}^{\perp}M_z^{out^{-1}}\widehat{P}\widehat{C}_0^{-1} + \mathcal{O}_{\mathcal{B}(\widehat{P}\mathbb{C}^N,\widehat{P}^{\perp}\mathbb{C}^N)}(\varepsilon) & \varepsilon^{-1}N_z + \mathcal{O}_{\mathcal{B}(\widehat{P}^{\perp}\mathbb{C}^N)}(1) \end{pmatrix}.
\end{aligned}$$

Hence Expansion (53) must hold true □

Recall that, for $\operatorname{Im} z \neq 0$, $\widehat{P}M_z^{out}\widehat{P}$ is invertible in $\widehat{P}\mathbb{C}^N$, see Remark A3.

Proposition 6. *Let $z \in \mathbb{C}\backslash\mathbb{R}$. In the non-generic case,*

$$(M_z^{in,\varepsilon}M_z^{out} - \mathbb{I}_N)^{-1}M_z^{in,\varepsilon} = \widehat{P}(\widehat{P}M_z^{out}\widehat{P})^{-1}\widehat{P} + \mathcal{O}_{\mathcal{B}(\mathbb{C}^N)}(\varepsilon).$$

Proof. Taking into account Expansion (55), rewritten in the decomposition $\mathbb{C}^N = \widehat{P}\mathbb{C}^N \oplus \widehat{P}^{\perp}\mathbb{C}^N$, one has

$$\widetilde{M}_z^{in,\varepsilon} = -\frac{1}{z}\widehat{C} + \mathcal{O}_{\mathcal{B}(\mathbb{C}^N)}(\varepsilon^2) = -\begin{pmatrix} z^{-1}\widehat{C}_0 & 0 \\ 0 & 0 \end{pmatrix} + \mathcal{O}_{\mathcal{B}(\mathbb{C}^N)}(\varepsilon^2).$$

So that, by Equation (53),

$$(\widetilde{M}_z^{in,\varepsilon} - \varepsilon M_z^{out^{-1}})^{-1}\widetilde{M}_z^{in,\varepsilon} = \begin{pmatrix} \mathbb{I}_{\widehat{P}\mathbb{C}^N} & 0 \\ -N_z\widehat{P}^{\perp}M_z^{out^{-1}}\widehat{P} & 0 \end{pmatrix} + \mathcal{O}_{\mathcal{B}(\mathbb{C}^N)}(\varepsilon).$$

By the latter expansion and by the identification (58) it follows that (recall Equation (51) and the definition of N_z in Equation (52))

$$\begin{aligned}
&(M_z^{in,\varepsilon} M_z^{out} - \mathbb{I}_N)^{-1} M_z^{in,\varepsilon} \\
=&M_z^{out^{-1}} (\widetilde{M}_z^{in,\varepsilon} - \varepsilon M_z^{out^{-1}})^{-1} \widetilde{M}_z^{in,\varepsilon} \\
=&\begin{pmatrix} \widehat{P} M_z^{out^{-1}} \widehat{P} & \widehat{P} M_z^{out^{-1}} \widehat{P}^\perp \\ \widehat{P}^\perp M_z^{out^{-1}} \widehat{P} & \widehat{P}^\perp M_z^{out^{-1}} \widehat{P}^\perp \end{pmatrix} \begin{pmatrix} \mathbb{I}_{\widehat{P}\mathbb{C}^N} & 0 \\ -N_z \widehat{P}^\perp M_z^{out^{-1}} \widehat{P} & 0 \end{pmatrix} + \mathcal{O}_{\mathcal{B}(\mathbb{C}^N)}(\varepsilon) \\
=&\begin{pmatrix} \widehat{P} M_z^{out^{-1}} \widehat{P} - \widehat{P} M_z^{out^{-1}} \widehat{P}^\perp N_z \widehat{P}^\perp M_z^{out^{-1}} \widehat{P} & 0 \\ 0 & 0 \end{pmatrix} + \mathcal{O}_{\mathcal{B}(\mathbb{C}^N)}(\varepsilon).
\end{aligned} \tag{59}$$

To conclude, we apply the block-matrix inversion formula to Equation (58) to obtain

$$M_z^{out} = \begin{pmatrix} \widetilde{D}_z & -\widetilde{D}_z \widehat{P} M_z^{out^{-1}} \widehat{P}^\perp N_z \\ -N_z \widehat{P}^\perp M_z^{out^{-1}} \widehat{P} \widetilde{D}_z & N_z + N_z \widehat{P}^\perp M_z^{out^{-1}} \widehat{P} \widetilde{D}_z \widehat{P} M_z^{out^{-1}} \widehat{P}^\perp N_z \end{pmatrix},$$

with

$$\widetilde{D}_z = (\widehat{P} M_z^{out^{-1}} \widehat{P} - \widehat{P} M_z^{out^{-1}} \widehat{P}^\perp N_z \widehat{P}^\perp M_z^{out^{-1}} \widehat{P})^{-1}.$$

Hence it must be

$$\widehat{P} M_z^{out} \widehat{P} = \widetilde{D}_z = (\widehat{P} M_z^{out^{-1}} \widehat{P} - \widehat{P} M_z^{out^{-1}} \widehat{P}^\perp N_z \widehat{P}^\perp M_z^{out^{-1}} \widehat{P})^{-1},$$

so that

$$(\widehat{P} M_z^{out} \widehat{P})^{-1} = \widehat{P} M_z^{out^{-1}} \widehat{P} - \widehat{P} M_z^{out^{-1}} \widehat{P}^\perp N_z \widehat{P}^\perp M_z^{out^{-1}} \widehat{P}.$$

This, together with Equation (59), allows us to infer the expansion

$$(M_z^{in,\varepsilon} M_z^{out} - \mathbb{I}_N)^{-1} M_z^{in,\varepsilon} = \begin{pmatrix} (\widehat{P} M_z^{out} \widehat{P})^{-1} & 0 \\ 0 & 0 \end{pmatrix} + \mathcal{O}_{\mathcal{B}(\mathbb{C}^N)}(\varepsilon) = \widehat{P} (\widehat{P} M_z^{out} \widehat{P})^{-1} \widehat{P} + \mathcal{O}_{\mathcal{B}(\mathbb{C}^N)}(\varepsilon)$$

and conclude the proof of the proposition. □

We are now ready to state and prove the main theorem for the non-generic case. In the statement of the theorem, we assume that $\operatorname{Ker} \widehat{C} \subset \mathbb{C}^N$, i.e., $\widehat{P} \neq 0$. In this way the quantity $(\underline{\hat{c}}_k, \widehat{C}_0^{-1} \underline{\hat{c}}_{k'})_{\mathbb{C}^N}$ is certainly well defined. We discuss the case $\operatorname{Ker} \widehat{C} = \mathbb{C}^N$ (i.e., $\widehat{P} = 0$) separately in the proof of point (ii) of Theorem 2 (after the proof of Theorem 4).

Theorem 4. *Let $z \in \mathbb{C}\backslash\mathbb{R}$. In the non-generic case assume that* $\operatorname{Ker} \widehat{C} \subset \mathbb{C}^N$, *then*

$$R_z^\varepsilon = \begin{pmatrix} \widehat{R}_z^{out} + \mathcal{O}_{\mathcal{B}(\mathcal{H}^{out})}(\varepsilon) & \mathcal{O}_{\mathcal{B}(\mathcal{H}^{in,\varepsilon},\mathcal{H}^{out})}(\varepsilon^{1/2}) \\ \mathcal{O}_{\mathcal{B}(\mathcal{H}^{out},\mathcal{H}^{in,\varepsilon})}(\varepsilon^{1/2}) & -z^{-1} \sum_{k,k'=1}^m \left(\delta_{k,k'} - (\underline{\hat{c}}_k, \widehat{C}_0^{-1} \underline{\hat{c}}_{k'})_{\mathbb{C}^N} \right) \varphi_k^\varepsilon (\varphi_{k'}^\varepsilon, \cdot)_{\mathcal{H}^{in,\varepsilon}} + \mathcal{O}_{\mathcal{B}(\mathcal{H}^{in,\varepsilon})}(\varepsilon). \end{pmatrix},$$

where the expansion has to be understood in the out/in decomposition (11).

Proof. We analyze term by term the r.h.s. in Equation (37).

Term *out*/*out*: by Proposition 6 and Lemma 1, it immediately follows that

$$\mathring{R}_z^{out} - \mathcal{R}_z^{out,out,\varepsilon} = \widehat{R}_z^{out} + \mathcal{O}_{\mathcal{B}(\mathcal{H}^{out})}(\varepsilon).$$

Term out/in: by Equation (51) and by the definition of $\mathcal{R}_z^{out,in,\varepsilon}$, recalling that G_z^{out} and M_z^{out-1} are bounded, it is enough to prove that

$$\varepsilon(\widetilde{M}_z^{in,\varepsilon} - \varepsilon M_z^{out-1})^{-1}\check{G}_z^{in,\varepsilon} = \mathcal{O}_{\mathcal{B}(\mathcal{H}^{in,\varepsilon},\mathbb{C}^N)}(\varepsilon^{1/2}). \tag{60}$$

Taking into account the fact that for all $\psi \in \mathcal{H}^{in,\varepsilon}$, $\|\sum_{k=1}^m \underline{\hat{c}}_k(\hat{\varphi}_k^\varepsilon, \psi)_{\mathcal{H}^{in,\varepsilon}}\|_{\mathbb{C}^N} \leq C\|\psi\|_{\mathcal{H}^{in,\varepsilon}}$, and the fact that $\sum_{k=1}^m \underline{\hat{c}}_k(\hat{\varphi}_k^\varepsilon, \psi)_{\mathcal{H}^{in,\varepsilon}} \in \widehat{P}\mathbb{C}^N$ (it is a linear combination of vectors in $\widehat{P}\mathbb{C}^N$, see Rem 4) we infer that (see Equation (47)),

$$\check{G}_z^{in,\varepsilon}\psi = \underline{q}^\varepsilon + \underline{p}^\varepsilon \qquad \underline{q}^\varepsilon := -\sum_{k=1}^m \frac{\underline{\hat{c}}_k(\hat{\varphi}_k^\varepsilon, \psi)_{\mathcal{H}^{in,\varepsilon}}}{\varepsilon^{1/2} z}$$

with $\underline{q}^\varepsilon \in \widehat{P}\mathbb{C}^N$, $\|\underline{q}^\varepsilon\|_{\mathbb{C}^N} \leq C\varepsilon^{-1/2}\|\psi\|_{\mathcal{H}^{in,\varepsilon}}$, and $\|\underline{p}^\varepsilon\|_{\mathbb{C}^N} \leq C\varepsilon^{3/2}\|\psi\|_{\mathcal{H}^{in,\varepsilon}}$.

Hence, by the expansion (53), we infer

$$\begin{aligned}
&\varepsilon(\widetilde{M}_z^{in,\varepsilon} - \varepsilon M_z^{out-1})^{-1}\check{G}_z^{in,\varepsilon}\psi \\
&= -\varepsilon\big(z\widehat{C}_0^{-1} - zN_z\widehat{P}^\perp M_z^{out-1}\widehat{P}\widehat{C}_0^{-1} + \mathcal{O}_{\mathcal{B}(\mathbb{C}^N)}(\varepsilon)\big)\underline{q}^\varepsilon + \varepsilon(\widetilde{M}_z^{in,\varepsilon} - \varepsilon M_z^{out-1})^{-1}\underline{p}^\varepsilon.
\end{aligned} \tag{61}$$

Here the leading term is

$$\varepsilon\big(z\widehat{C}_0^{-1} - zN_z\widehat{P}^\perp M_z^{out-1}\widehat{P}\widehat{C}_0^{-1}\big)\underline{q}^\varepsilon,$$

and for it we have the bound

$$\|\varepsilon\big(z\widehat{C}_0^{-1} - zN_z\widehat{P}^\perp M_z^{out-1}\widehat{P}\widehat{C}_0^{-1}\big)\underline{q}^\varepsilon\|_{\mathbb{C}^N} \leq C\varepsilon^{1/2}\|\psi\|_{\mathcal{H}^{in,\varepsilon}}.$$

The remainder is bounded by

$$\|\mathcal{O}_{\mathcal{B}(\mathbb{C}^N)}(\varepsilon^2)\underline{q}^\varepsilon + \varepsilon(\widetilde{M}_z^{in,\varepsilon} - \varepsilon M_z^{out-1})^{-1}\underline{p}^\varepsilon\|_{\mathbb{C}^N} \leq C\varepsilon^2\|\underline{q}^\varepsilon\|_{\mathbb{C}^N} + C\|\underline{p}^\varepsilon\|_{\mathbb{C}^N} \leq C\varepsilon^{3/2}\|\psi\|_{\mathcal{H}^{in,\varepsilon}};$$

in the latter bound we used $(\widetilde{M}_z^{in,\varepsilon} - \varepsilon M_z^{out-1})^{-1} = \mathcal{O}_{\mathcal{B}(\mathbb{C}^N)}(\varepsilon^{-1})$, see Equation (53) (see also Equation (54)). Hence,

$$\|\varepsilon(\widetilde{M}_z^{in,\varepsilon} - \varepsilon M_z^{out-1})^{-1}\check{G}_z^{in,\varepsilon}\psi\|_{\mathbb{C}^N} \leq C\varepsilon^{1/2}\|\psi\|_{\mathcal{H}^{in,\varepsilon}},$$

and the bound (60) holds true.

The bound on the term in/out follows immediately by noticing that $\mathcal{R}_z^{in,out,\varepsilon} = \mathcal{R}_{\bar{z}}^{out,in,\varepsilon*}$.

Term in/in; by Equation (51), we have that

$$\mathcal{R}_z^{in,in,\varepsilon} = \varepsilon G_z^{in,\varepsilon}(\widetilde{M}_z^{in,\varepsilon} - \varepsilon M_z^{out-1})^{-1}\check{G}_z^{in,\varepsilon}.$$

Taking into account Equation (61) and the expansion (48), we infer that, for all $\psi \in \mathcal{H}^{in,\varepsilon}$ the leading term in $\mathcal{R}_z^{in,in,\varepsilon}\psi$ is given by

$$\begin{aligned}
\sum_{k=1}^m \frac{\hat{\varphi}_k^\varepsilon(\underline{\hat{c}}_k, \cdot)_{\mathbb{C}^N}}{\varepsilon^{1/2} z}\big(\varepsilon\big(z\widehat{C}_0^{-1} - zN_z\widehat{P}^\perp M_z^{out-1}\widehat{P}\widehat{C}_0^{-1}\big)\underline{q}^\varepsilon\big) =& \varepsilon^{1/2}\sum_{k=1}^m \hat{\varphi}_k^\varepsilon(\underline{\hat{c}}_k, \widehat{C}_0^{-1}\underline{q}^\varepsilon)_{\mathbb{C}^N} \\
=& -\frac{1}{z}\sum_{k,k'=1}^m \hat{\varphi}_k^\varepsilon(\underline{\hat{c}}_k, \widehat{C}_0^{-1}\underline{\hat{c}}_{k'})_{\mathbb{C}^N}(\hat{\varphi}_{k'}^\varepsilon, \psi)_{\mathcal{H}^{in,\varepsilon}}.
\end{aligned}$$

the remainder being of order ε. From the latter formula and from the expansion (46) we infer

$$\mathring{R}_z^{in,\varepsilon} - \mathcal{R}_z^{in,in,\varepsilon} = -z^{-1}\sum_{k,k'=1}^m \Big(\delta_{k,k'} - (\underline{\hat{c}}_k, \widehat{C}_0^{-1}\underline{\hat{c}}_{k'})_{\mathbb{C}^N}\Big)\hat{\varphi}_k^\varepsilon(\hat{\varphi}_{k'}^\varepsilon, \cdot)_{\mathcal{H}^{in,\varepsilon}} + \mathcal{O}_{\mathcal{B}(\mathcal{H}^{in,\varepsilon})}(\varepsilon).$$

□

Theorem 2-(i) follows immediately from Theorem 4.

Proof of Theorem 2 - (ii). If $\operatorname{Ker}\widehat{C} = \mathbb{C}^N$ then $\underline{\hat{c}}_k = 0$, for all $k = 1, \dots, m$, see Remark 4. Hence, expansions (47), (48), and (50) read respectively

$$\check{G}_z^{in,\varepsilon} = \mathcal{O}_{\mathcal{B}(\mathcal{H}^{in,\varepsilon},\mathbb{C}^N)}(\varepsilon^{3/2}); \qquad G_z^{in,\varepsilon} = \mathcal{O}_{\mathcal{B}(\mathbb{C}^N,\mathcal{H}^{in,\varepsilon})}(\varepsilon^{3/2}); \qquad M_z^{in,\varepsilon} = \mathcal{O}_{\mathcal{B}(\mathbb{C}^N)}(\varepsilon).$$

Reasoning along the lines of the analysis of the generic case, see the proof of Theorem 3, and taking into account the expansion (46), one readily infers

$$R_z^\varepsilon = \begin{pmatrix} \mathring{R}_z^{out} + \mathcal{O}_{\mathcal{B}(\mathcal{H}^{out})}(\varepsilon) & \mathcal{O}_{\mathcal{B}(\mathcal{H}^{in,\varepsilon},\mathcal{H}^{out})}(\varepsilon^{3/2}) \\ \mathcal{O}_{\mathcal{B}(\mathcal{H}^{out},\mathcal{H}^{in,\varepsilon})}(\varepsilon^{3/2}) & -\sum_{k=1}^m \frac{\varphi_k^\varepsilon(\varphi_k^\varepsilon,\cdot)_{\mathcal{H}^{in,\varepsilon}}}{z} + \mathcal{O}_{\mathcal{B}(\mathcal{H}^{in,\varepsilon})}(\varepsilon^2), \end{pmatrix}$$

which implies the statement in Theorem 2 - (ii). □

Proof of Theorem 2 - (iii). To prove the second part of Theorem 2, recall that $\underline{\hat{c}}_{k'} \in \widehat{P}\mathbb{C}^N$ and $\widehat{C}_0^{-1}\underline{\hat{c}}_{k'} \in \widehat{P}\mathbb{C}^N$, hence $\widehat{C}\widehat{C}_0^{-1}\underline{\hat{c}}_{k'} = \widehat{C}_0\widehat{C}_0^{-1}\underline{\hat{c}}_{k'} = \underline{\hat{c}}_{k'}$. By the definition of $\widehat{C}$ this is equivalent to

$$\sum_{k=1}^m (\delta_{k,k'} - (\underline{\hat{c}}_k, \widehat{C}_0^{-1}\underline{\hat{c}}_{k'}))\underline{\hat{c}}_k = 0.$$

If the vectors $\{\underline{\hat{c}}_k\}_{k=1}^m$ are linearly independent this linear combination is zero if and only if $\delta_{k,k'} - (\underline{\hat{c}}_k, \widehat{C}_0^{-1}\underline{\hat{c}}_{k'}) = 0$ for all k. Hence, expansion (15) follows from Equation (14). □

Remark 9. *Denote by Λ the operator in $\mathcal{H}^{in,\varepsilon}$ defined by*

$$D(\Lambda) := \mathcal{H}^{in,\varepsilon}, \qquad \Lambda := \sum_{k,k'=1}^m \left(\delta_{k,k'} - (\underline{\hat{c}}_k, \widehat{C}_0^{-1}\underline{\hat{c}}_{k'})_{\mathbb{C}^N}\right) \varphi_k^\varepsilon(\varphi_{k'}^\varepsilon, \cdot)_{\mathcal{H}^{in,\varepsilon}}.$$

Λ is selfadjoint and $\Lambda^2 = \Lambda$. The first claim is obvious (recall that $\widehat{C}_0$ is selfadjoint). To prove the second claim, note that, since $(\varphi_{l'}^\varepsilon, \varphi_k^\varepsilon)_{\mathcal{H}^{in,\varepsilon}} = \delta_{l',k}$,

$$\Lambda^2 = \sum_{l,k,k'=1}^m \left(\delta_{l,k} - (\underline{\hat{c}}_l, \widehat{C}_0^{-1}\underline{\hat{c}}_k)_{\mathbb{C}^N}\right)\left(\delta_{k,k'} - (\underline{\hat{c}}_k, \widehat{C}_0^{-1}\underline{\hat{c}}_{k'})_{\mathbb{C}^N}\right) \varphi_l^\varepsilon(\varphi_{k'}^\varepsilon, \cdot)_{\mathcal{H}^{in,\varepsilon}},$$

but

$$\begin{aligned} &\sum_{k=1}^m \left(\delta_{l,k} - (\underline{\hat{c}}_l, \widehat{C}_0^{-1}\underline{\hat{c}}_k)_{\mathbb{C}^N}\right)\left(\delta_{k,k'} - (\underline{\hat{c}}_k, \widehat{C}_0^{-1}\underline{\hat{c}}_{k'})_{\mathbb{C}^N}\right) \\ =&\delta_{l,k'} - 2(\underline{\hat{c}}_l, \widehat{C}_0^{-1}\underline{\hat{c}}_{k'})_{\mathbb{C}^N} + \sum_{k=1}^m (\underline{\hat{c}}_l, \widehat{C}_0^{-1}\underline{\hat{c}}_k)_{\mathbb{C}^N}(\underline{\hat{c}}_k, \widehat{C}_0^{-1}\underline{\hat{c}}_{k'})_{\mathbb{C}^N} \\ =&\delta_{l,k'} - 2(\underline{\hat{c}}_l, \widehat{C}_0^{-1}\underline{\hat{c}}_{k'})_{\mathbb{C}^N} + (\underline{\hat{c}}_l, \widehat{C}_0^{-1}\widehat{C}\widehat{C}_0^{-1}\underline{\hat{c}}_{k'})_{\mathbb{C}^N} = \delta_{l,k'} - (\underline{\hat{c}}_l, \widehat{C}_0^{-1}\underline{\hat{c}}_{k'})_{\mathbb{C}^N}, \end{aligned}$$

where we used the fact that $\widehat{C}_0^{-1}\widehat{C}\widehat{C}_0^{-1} = \widehat{C}_0^{-1}\widehat{C}_0\widehat{C}_0^{-1} = \widehat{C}_0^{-1}$. Hence,

$$\Lambda^2 = \sum_{l,k'=1}^m \left(\delta_{l,k'} - (\underline{\hat{c}}_l, \widehat{C}_0^{-1}\underline{\hat{c}}_{k'})_{\mathbb{C}^N}\right) \varphi_l^\varepsilon(\varphi_{k'}^\varepsilon, \cdot)_{\mathcal{H}^{in,\varepsilon}} = \Lambda.$$

Hence, Λ is an orthogonal projection in $\mathcal{H}^{in,\varepsilon}$.

Funding: This research received no external funding.

Acknowledgments: The author is grateful to Gregory Berkolaiko and Andrea Posilicano for enlightening discussions. The author also thanks the anonymous referees for many useful comments that helped to improve the quality of the paper.

Conflicts of Interest: The authors declare no conflict of interest.

Appendix A. Proof of the Kreĭn Resolvent Formulae

We use several known results from the theory of self-adjoint extensions of symmetric operators. We follow, for the most, the approach and the notation from the papers by A. Posilicano [24] and [22]. Other approaches would be possible, such as the one based on the use of boundary triples, see, e.g., [25–28].

When no misunderstanding is possible, in this appendix we omit the suffixes "*out*", "*in*", and ε.

Appendix A.1. Proofs of Lemmata 1 and 2

We denote by $\mathring{\tau}$ the restriction of the maps τ to the domain $D(\mathring{H})$, by Equations (22) and (23) we infer

$$\mathring{\tau} : D(\mathring{H}^\varepsilon) \to \mathbb{C}^{2N}, \qquad \mathring{\tau} = \operatorname{diag}(\mathring{\tau}^{out}, \mathring{\tau}^{in});$$
$$\mathring{\tau}^{out} : D(\mathring{H}^{out}) \to \mathbb{C}^{N}, \qquad \mathring{\tau}^{out}\psi := \Psi'(\mathbf{0});$$
$$\mathring{\tau}^{in} : D(\mathring{H}^{in,\varepsilon}) \to \mathbb{C}^{N}, \qquad \mathring{\tau}^{in}\psi := (\psi(v_1), ..., \psi(v_N))^T;$$

where in $\mathring{\tau}^{in}$ we used the definition of τ^{in} and the fact that functions in $D(\mathring{H}^{in,\varepsilon})$ are continuous in the connecting vertices.

Remark A1. *The map $\mathring{\tau}$ is surjective. Hence, the map $\check{G}^\varepsilon_z = \tau\mathring{R}^\varepsilon_z = \mathring{\tau}\mathring{R}^\varepsilon_z$ is also surjective as a map from $\mathcal{H}^\varepsilon \to \mathbb{C}^{2N}$ (the operator $\mathring{R}^\varepsilon_z : \mathcal{H}^\varepsilon \to D(\mathring{H}^\varepsilon)$ is obviously surjective). We conclude that $G^\varepsilon_z = \check{G}^{\varepsilon *}_{\bar{z}}$ is an injective map from $\mathbb{C}^{2N} \to \mathcal{H}^\varepsilon$ (it is the adjoint of a surjective map). A similar statement holds true also for the corresponding "out" and "in" operators.*

Remark A2. *We claim that for all $z \in \rho(\mathring{H}^\varepsilon)$ and $\underline{q} \in \mathbb{C}^{2N}$ one has $G^\varepsilon_z\underline{q} \in \mathcal{H}^\varepsilon_2$ and*

$$(-\Delta + B^\varepsilon - z)G^\varepsilon_z\underline{q} = 0, \tag{A1}$$

and similar properties hold true for the "out" and "in" operators (here Δ denotes the maximal Laplacian in $\mathcal{H}^\varepsilon$, i.e., $D(\Delta) := \mathcal{H}^\varepsilon_2$, $\Delta\psi = \psi''$).

To prove that $G^\varepsilon_z\underline{q} \in \mathcal{H}^\varepsilon_2$ and that Equation (A1) holds true we start by discussing the case $B^\varepsilon = 0$. In such a case it is possible to obtain an explicit formula for the integral kernel of $\mathring{R}^\varepsilon_{z,0} = \mathring{R}^\varepsilon_{z,B^\varepsilon=0}$, see, e.g., ([2], Lemma 4.2). By this explicit formula it is easily seen that the operator $G^\varepsilon_{z,0} = G^\varepsilon_{z,B^\varepsilon=0}$ maps any vector $\underline{q} \in \mathbb{C}^{2N}$ in a function in $\mathcal{H}^\varepsilon_2$ and that $(-\Delta - z)G^\varepsilon_{z,0}\underline{q} = 0$. It is not needed to investigate the detailed properties of the boundary conditions in the vertices of $\mathcal{G}^\varepsilon$, it is enough to take into account the dependence on $x, y \in \mathcal{G}^\varepsilon$ of the integral kernel $\mathring{R}^\varepsilon_{z,0}(x,y)$ (see also ([22], Examples 5.1 and 5.2)). That the same is true for $B^\varepsilon \neq 0$ follows immediately from the resolvent identity

$$\mathring{R}^\varepsilon_z = \mathring{R}^\varepsilon_{z,0} - \mathring{R}^\varepsilon_{z,0}B^\varepsilon\mathring{R}^\varepsilon_z,$$

which gives $\check{G}^\varepsilon_z = \check{G}^\varepsilon_{z,0} - \check{G}^\varepsilon_{z,0}B^\varepsilon\mathring{R}^\varepsilon_z$ and $G^\varepsilon_z = G^\varepsilon_{z,0} - \mathring{R}^\varepsilon_z B^\varepsilon G^\varepsilon_{z,0}$.

In consideration of the remark above, we infer that the maps ($N \times N$, z-dependent matrices) M_z in Equations (26), (27) and (28) are all well defined. Moreover, by the resolvent identities

$$R_z - R_w = (z-w)R_zR_w \qquad \text{and} \qquad R_z = R^*_{\bar{z}}$$

it follows that

$$\check{G}_z - \check{G}_w = (z-w)\check{G}_zR_w,$$

$$G_z - G_w = (z-w)R_w G_z,$$
$$M_z - M_w = (z-w)\check{G}_w G_z \qquad \text{and} \qquad M_z = M_{\bar{z}}^*. \tag{A2}$$

Let us denote by $\mathcal{K}$ the space $\mathbb{C}^{2N}$ or $\mathbb{C}^N$ depending on if we are reasoning with operators in $\mathcal{H}^\varepsilon$, $\mathcal{H}^{out}$ or $\mathcal{H}^{in,\varepsilon}$. By Equation (A2), it follows that for any projection P in $\mathcal{K}$ and any self-adjoint operator Θ in $\operatorname{Ran} P$, the map $M_z^{P,\Theta} := PM_zP - \Theta$ is invertible in $\operatorname{Ran} P$. To see that this is indeed the case, note that by Equation (A2) one has

$$M_z^{P,\Theta} - M_w^{P,\Theta} = (z-w)P\check{G}_w G_z P \qquad \text{and} \qquad M_z^{P,\Theta} = M_{\bar{z}}^{P,\Theta *}.$$

So that, for $\operatorname{Im} z \neq 0$ and for all $\underline{q} \in \mathcal{K}$, such that $P\underline{q} \neq 0$, it holds

$$\operatorname{Im}(\underline{q}, M_z^{P,\Theta}\underline{q})_{\mathcal{K}} = \frac{1}{2i}(\underline{q}, (M_z^{P,\Theta} - M_{\bar{z}}^{P,\Theta})\underline{q})_{\mathcal{K}} = \operatorname{Im} z \|G_z P\underline{q}\|_{\mathcal{H}}^2 \neq 0; \tag{A3}$$

because G_z is injective. Hence, $M_z^{P,\Theta}$ is invertible in $\operatorname{Ran} P$ for $\operatorname{Im} z \neq 0$.

Remark A3. *By the discussion above, it follows that the maps* $M_z^{out} : \mathbb{C}^N \to \mathbb{C}^N$, $\widehat{P} M_z^{out} \widehat{P} : \widehat{P}\mathbb{C}^N \to \widehat{P}\mathbb{C}^N$, *and* $(M_z^\varepsilon - \boldsymbol{\Theta}) : \mathbb{C}^{2N} \to \mathbb{C}^{2N}$ *are invertible for all* $\operatorname{Im} z \neq 0$.

By ([22], Theorem 2.1) (see also ([24], Theorem 2.1)) it follows that: for any $z \in \mathbb{C}\backslash\mathbb{R}$ the operators $\widehat{R}_z^{out}$ and R_z^ε are the resolvents of a self-adjoint extension of the symmetric operators $\mathring{H}^{out} \restriction_{\operatorname{Ker} \mathring{\tau}^{out}}$ and $\mathring{H}^\varepsilon \restriction_{\operatorname{Ker} \mathring{\tau}}$ respectively.

We are left to prove that such self-adjoint extensions coincide with $\widehat{H}^{out}$ and H^ε respectively.

Let us focus attention on R_z^ε (similar considerations hold true for $\widehat{R}_z^{out}$). Since the self-adjoint operator associated to R_z^ε is an extension of $\mathring{H}^\varepsilon \restriction_{\operatorname{Ker} \mathring{\tau}}$, to prove that R_z^ε is the resolvent of H^ε, we just need to check that in the connecting vertices functions in $\operatorname{Ran} R_z^\varepsilon$ satisfy the boundary conditions required by $D(H^\varepsilon)$. The remaining boundary conditions are clearly satisfied because the map $\mathring{\tau}$ evaluates functions only in the connecting vertices.

Define the maps:

$$\sigma^{out} : \mathcal{H}_2^{out} \to \mathbb{C}^N \qquad \sigma^{out}\psi := \Psi(\mathbf{0});$$

$$\sigma^{in} : \mathcal{H}_2^{in,\varepsilon} \to \mathbb{C}^N$$
$$\sigma^{in}\psi := -\left(\sqrt{d^{in}(v_1)}(\mathbf{1}_{d^{in}(v_1)}, \Psi'(v_1))_{\mathbb{C}^{d^{in}(v_1)}}, \dots, \sqrt{d^{in}(v_N)}(\mathbf{1}_{d^{in}(v_N)}, \Psi'(v_N))_{\mathbb{C}^{d^{in}(v_N)}}\right)^T;$$

and

$$\sigma : \mathcal{H}_2^\varepsilon = \mathcal{H}_2^{out} \oplus \mathcal{H}_2^{in,\varepsilon} \to \mathbb{C}^{2N} \qquad \sigma := \operatorname{diag}(\sigma^{out}, \sigma^{in}).$$

We recall the following formula which is obtained by integrating by parts

$$((-\Delta + B^\varepsilon - \bar{z})\phi, \psi)_{\mathcal{H}^\varepsilon} - (\phi, (-\Delta + B^\varepsilon - z)\psi)_{\mathcal{H}^\varepsilon} = \sum_{v \in \mathcal{V}} \left[(\Phi'(v), \Psi(v))_{\mathbb{C}^{d(v)}} - (\Phi(v), \Psi'(v))_{\mathbb{C}^{d(v)}}\right] \tag{A4}$$
$$\forall \phi, \psi \in \mathcal{H}_2^\varepsilon.$$

Fix $\chi \in \mathcal{H}^\varepsilon$ and let $\underline{q} = \left(M_z^\varepsilon - \boldsymbol{\Theta}\right)^{-1} \check{G}_z^\varepsilon \chi \in \mathbb{C}^{2N}$ and $\psi = G_z^\varepsilon \underline{q}$.

For all $\phi \in D(\mathring{H}^\varepsilon)$ and ψ as above, the identity (A4) gives

$$(\tau\phi, \underline{q})_{\mathbb{C}^{2N}} = \Sigma_{v \in \mathcal{C}}\left[(K_v^{in\perp}\Phi^{in\prime}(v), K_v^{in\perp}\Psi^{in}(v))_{\mathbb{C}^{d(v)}} - (K_v^{in}\Phi^{in}(v), K_v^{in}\Psi^{in\prime}(v))_{\mathbb{C}^{d(v)}}\right] + \Sigma_{j=1}^N \overline{\phi_j^{out}}'(0)\psi_j^{out}(0). \tag{A5}$$

In what follows we use the decomposition $\mathbb{C}^{2N} = \mathbb{C}^N \oplus \mathbb{C}^N$, so that $\underline{q} = (\underline{q}^{out}, \underline{q}^{in})$ and $\tau\phi = (\tau^{out}\phi^{out}, \tau^{in}\phi^{in})$.

Let $\phi = (\phi^{out}, 0) \in D(\mathring{H}^\varepsilon)$. Then Identity (A5) gives

$$(\tau^{out}\phi^{out}, \underline{q}^{out})_{\mathbb{C}^N} = \sum_{j=1}^{N} \overline{\phi_j^{out\prime}}(0)\psi_j^{out}(0). \tag{A6}$$

Take $\phi^{out} \in D(\mathring{H}^{out})$, such that $\phi_1^{out\prime}(0) = 1$ and $\phi_j^{out} = 0$ for all $j = 2, \ldots, N$. Then $(\tau^{out}\phi^{out})_j = \delta_{1,j}$, $j = 1, \ldots, N$ and Equation (A6) gives $\psi_1^{out}(0) = q_1$. In a similar way it is possible to show that $\psi_j^{out}(0) = q_j$ for all $j = 2, \ldots, N$. Hence, $\sigma^{out}\psi^{out} = \underline{q}^{out}$.

Next let $\phi = (0, \phi^{in})$. Then Identity (A5) gives

$$(\tau^{in}\phi^{in}, \underline{q}^{in})_{\mathbb{C}^N} = \sum_{v \in \mathcal{C}} \left[(K_v^{in\perp}\Phi^{in\prime}(v), K_v^{in\perp}\Psi^{in}(v))_{\mathbb{C}^{d(v)}} - (K_v^{in}\Phi^{in}(v), K_v^{in}\Psi^{in\prime}(v))_{\mathbb{C}^{d(v)}} \right]. \tag{A7}$$

Take ϕ^{in} such that $\phi^{in}(v_1) = 1$, $\Phi^{in\prime}(v_1) = 0$ and $\Phi^{in\prime}(v_j) = \Phi^{in}(v_j) = 0$ for all $j = 2, \ldots, N$. Hence, $(\tau^{in}\phi^{in})_j = \delta_{1,j}$, $j = 1, \ldots, N$, and $K_{v_1}^{in}\Phi^{in}(v_1) = (d^{in}(v_1))^{1/2}\mathbf{1}_{d^{in}(v_1)}$. Hence, Equation (A7) gives

$$q_1^{in} = -\left((d^{in}(v_1))^{1/2}\mathbf{1}_{d^{in}(v_1)}, K_{v_1}^{in}\Psi^{in\prime}(v_1)\right)_{\mathbb{C}^{d(v_1)}} = -\left((d^{in}(v_1))^{1/2}\mathbf{1}_{d^{in}(v_1)}, \Psi^{in\prime}(v_1)\right)_{\mathbb{C}^{d(v_1)}} = (\sigma^{in}\psi^{in})_1.$$

In a similar way one can prove $q_j^{in} = (\sigma^{in}\psi^{in})_j$, $j = 2, \ldots, N$, hence, $\sigma^{in}\psi^{in} = \underline{q}^{in}$.

We also note that the function ψ is continuous in the connecting vertices (whenever the vertex degree is larger or equal than two). To see that this is indeed the case, consider in Equation (A7) a function ϕ^{in} such that $\phi^{in}(v_j) = 0$, $j = 1, \ldots, N$, $\Phi^{in\prime}(v_1) = (1, -1, 0, \ldots, 0)^T := \underline{e}$, $\Phi^{in\prime}(v_j) = 0$, $j = 2, \ldots, N$. Since $K_{v_1}^{in\perp}\underline{e} = \underline{e}$, condition (A7) gives $(\underline{e}, \Psi^{in}(v_1)) = 0$. Repeating the process, moving -1 in the vector $\underline{e}$ on all the positions (from the second one on) one obtains the continuity of ψ in the vertex v_1. The same holds true for every connecting vertex.

We have proved that for any $\chi \in \mathcal{H}^\varepsilon$, setting $\underline{q} = \left(M_z^\varepsilon - \mathbf{\Theta}\right)^{-1}\check{G}_z^\varepsilon\chi \in \mathbb{C}^{2N}$, one has:

$$\sigma^{out}G_z^{out}\underline{q}^{out} = \underline{q}^{out}; \qquad \sigma^{in}G_z^{in,\varepsilon}\underline{q}^{in} = \underline{q}^{in}; \qquad \sigma G_z^\varepsilon\underline{q} = \underline{q}. \tag{A8}$$

Let $\chi \in \mathcal{H}^\varepsilon$ and set $\psi = R_z^\varepsilon\chi$. One has that

$$\tau\psi = \tau\left(\mathring{R}_z^\varepsilon - G_z^\varepsilon\left(M_z^\varepsilon - \mathbf{\Theta}\right)^{-1}\check{G}_z^\varepsilon\right)\chi = \left(\mathbb{I} - M_z^\varepsilon(M_z^\varepsilon - \mathbf{\Theta})^{-1}\right)\check{G}_z^\varepsilon\chi = -\mathbf{\Theta}(M_z^\varepsilon - \mathbf{\Theta})^{-1}\check{G}_z^\varepsilon\chi.$$

On the other hand, noticing that $\sigma\mathring{R}_z^\varepsilon\chi = 0$, by the definition of $D(\mathring{H}^\varepsilon)$ (see Equations (10), (6), and (17)), and by Equation (A8) it follows that

$$\sigma\psi = -(M_z^\varepsilon - \mathbf{\Theta})^{-1}\check{G}_z^\varepsilon\chi.$$

We conclude that ψ satisfies the condition $\tau\psi = \mathbf{\Theta}\sigma\psi$. Taking into account the fact that ψ^{in} is continuous in the connecting vertices, it is easy convince oneself that the condition $\tau\psi = \mathbf{\Theta}\sigma\psi$ is equivalent to

$$\Psi^{out\prime}(0) = -\left(\sqrt{d^{in}(v_1)}(\mathbf{1}_{d^{in}(v_1)}, \Psi^{in\prime}(v_1))_{\mathbb{C}^{d^{in}(v_1)}}, \ldots, \sqrt{d^{in}(v_N)}(\mathbf{1}_{d^{in}(v_N)}, \Psi^{in\prime}(v_N))_{\mathbb{C}^{d^{in}(v_N)}}\right)^T,$$

and

$$\psi^{in}(v_j) = \psi_j(0);$$

which, in turns, is equivalent to the Kirchhoff boundary conditions in $D(H^\varepsilon)$.

The fact that the resolvent formula holds true for all $z \in \rho(H^\varepsilon) \cap \rho(\mathring{H}^\varepsilon)$, follows from ([29], Theorem 2.19).

To prove the resolvent formula for $\widehat{R}_z^{out}$, let $\chi \in \mathcal{H}^{out}$ and set $\psi = \widehat{R}_z^{out}\chi$. By the first formula in (A8), one has

$$\Psi(0) = -\widehat{P}(\widehat{P}M^{out}\widehat{P})^{-1}\widehat{P}\check{G}_z^{out}\chi,$$

hence, $\widehat{P}^{\perp}\Psi(\mathbf{0}) = 0$. Moreover,

$$\widehat{P}\Psi'(\mathbf{0}) = \widehat{P}\tau^{out}\psi = (\mathbb{I} - \widehat{P}M_z^{out}\widehat{P}(\widehat{P}M^{out}\widehat{P})^{-1})\widehat{P}\check{G}_z^{out}\chi = 0.$$

Hence, the boundary conditions in $D(\widehat{H}^{out})$ are satisfied, see Definition 7.

Appendix A.2. Proof of Lemma 3

Recall that we are denoting by $\mathcal{K}$ the space $\mathbb{C}^{2N}$ or $\mathbb{C}^N$ depending on if we are reasoning with operators in $\mathcal{H}^{\varepsilon}$, $\mathcal{H}^{out}$ or $\mathcal{H}^{in,\varepsilon}$.

Remark A4. *By Identities (A2) we infer*

$$M_w^{-1} - M_z^{-1} = (z-w)M_w^{-1}\check{G}_w G_z M_z^{-1}.$$

Hence, for $\operatorname{Im} z \neq 0$, *and for any projection P in $\mathcal{K}$, and $\underline{q} \in P\mathcal{K}$*

$$\operatorname{Im}(\underline{q}, PM_z^{-1}P\underline{q})_{\mathcal{K}} = \frac{1}{2i}(\underline{q}, P(M_z^{-1} - M_{\bar{z}}^{-1})P\underline{q})_{\mathcal{K}} = -\operatorname{Im} z\|G_z M_z^{-1}P\underline{q}\|_{\mathcal{H}}^2 \neq 0 \tag{A9}$$

because $G_z M_z^{-1}$ is an injective map, being the composition of injective maps.

Hence, the map $PM_z^{-1}P$ is invertible in $P\mathcal{K}$.

To prove that the map $M_z^{in,\varepsilon}M_z^{out} - \mathbb{I}_N$ is invertible (the proof of the second statement in Equation (30) is analogous) note that it is enough to show that $M_z^{in,\varepsilon} - M_z^{out^{-1}}$ is invertible (because M_z^{out} is). Let $\underline{q} \in \mathbb{C}^N$, by Equations (A3) and (A9)

$$\operatorname{Im}(\underline{q}, M_z^{in,\varepsilon} - M_z^{out^{-1}}\underline{q})_{\mathbb{C}^N} = \operatorname{Im} z(\|G_z^{in,\varepsilon}\underline{q}\|_{\mathbb{C}^N}^2 + \|G_z^{out}M_z^{out^{-1}}\underline{q}\|_{\mathbb{C}^N}^2) \neq 0.$$

Formula (31), comes from the block matrix inversion formula

$$\begin{pmatrix} M_z^{out} & -\mathbb{I}_N \\ -\mathbb{I}_N & M_z^{in,\varepsilon} \end{pmatrix}^{-1} = \begin{pmatrix} M_z^{out^{-1}} + M_z^{out^{-1}}(M_z^{in,\varepsilon} - M_z^{out^{-1}})^{-1}M_z^{out^{-1}} & M_z^{out^{-1}}(M_z^{in,\varepsilon} - M_z^{out^{-1}})^{-1} \\ (M_z^{in,\varepsilon} - M_z^{out^{-1}})^{-1}M_z^{out^{-1}} & (M_z^{in,\varepsilon} - M_z^{out^{-1}})^{-1} \end{pmatrix},$$

together with the identities

$$M_z^{out^{-1}}(M_z^{in,\varepsilon} - M_z^{out^{-1}})^{-1} = (M_z^{in,\varepsilon}M_z^{out} - \mathbb{I}_N)^{-1}$$
$$(M_z^{in,\varepsilon} - M_z^{out^{-1}})^{-1}M_z^{out^{-1}} = (M_z^{out}M_z^{in,\varepsilon} - \mathbb{I}_N)^{-1}$$

and

$$M_z^{out^{-1}} + M_z^{out^{-1}}(M_z^{in,\varepsilon} - M_z^{out^{-1}})^{-1}M_z^{out^{-1}} = (M_z^{in,\varepsilon}M_z^{out} - \mathbb{I}_N)^{-1}M_z^{in,\varepsilon}.$$

Appendix B. Estimates on Eigenvalues and Eigenfunctions of $\mathring{H}^{in}$

In this appendix we prove the following proposition on the asymptotic behavior of eigenvalues and eigenfunctions of $\mathring{H}^{in}$.

Proposition A1. *Recall that we denoted by* $\{\lambda_n\}_{n\in\mathbb{N}}$ *the eigenvalues of the Hamiltonian* $\mathring{H}^{in}$*, and by* $\{\varphi_n\}_{n\in\mathbb{N}}$ *a corresponding set of orthonormal eigenfunctions. There exists* n_0 *such that for any* $n \geq n_0$*:*

$$\lambda_n > n^2 C \tag{A10}$$

and

$$\sup_{x\in\mathcal{G}^{in}} |\varphi_n(x)| \leq C \tag{A11}$$

for some positive constant C which does not depend on n.

Proof. Claim (A10) is just the Weyl law. For $B^{in} = 0$ a proof can be found in ([30], Proposition 4.2) (see also [31]). For $B^{in} \neq 0$ bounded, claim (A10) can be deduced by a perturbative argument.

To prove the bound (A11) we follow the lines in the proof of Theorem A.1 in [32]. For $b \in L^\infty(0,\ell)$ and real valued, and $\lambda > 0$ let f be the solution of the equation

$$-f'' + bf = \lambda f, \tag{A12}$$

with initial conditions $f(0) = f_0$ and $f'(0) = f_0'$. Then $f(x)$ can be written as

$$f(x) = \int_0^x \frac{\sin(\sqrt{\lambda}(x-y))}{\sqrt{\lambda}} b(y)f(y)dy + f_0\cos(\sqrt{\lambda}x) + \frac{f_0'}{\sqrt{\lambda}}\sin(\sqrt{\lambda}x), \tag{A13}$$

from which it immediately follows that

$$|f(x)| \leq M + \int_0^x \frac{1}{\sqrt{\lambda}}|b(y)||f(y)|dy,$$

with

$$M = |f_0| + \frac{|f_0'|}{\sqrt{\lambda}}.$$

Then from Gronwall's lemma, see, e.g. ([33], page 103), one has

$$|f(x)| \leq M\exp\left(\int_0^x \frac{|b(y)|}{\sqrt{\lambda}}dy\right) \leq M\exp\left(\int_0^\ell |b(y)|dy\right), \tag{A14}$$

where we assumed $\lambda > 1$. By equation (A13) and by the estimate (A14) it follows that

$$\left|f(x) - f_0\cos(\sqrt{\lambda}x) - \frac{f_0'}{\sqrt{\lambda}}\sin(\sqrt{\lambda}x)\right| \leq M\exp\left(\int_0^\ell |b(y)|dy\right)\int_0^x \frac{|b(y)|}{\sqrt{\lambda}}dy \leq C\left(\frac{|f_0|}{\sqrt{\lambda}} + \frac{|f_0'|}{\lambda}\right)$$

where C is a positive constant which does not depend on λ, f_0 and f_0'. We have then proved that

$$f(x) = f_0\cos(\sqrt{\lambda}x) + \frac{f_0'}{\sqrt{\lambda}}\sin(\sqrt{\lambda}x) + \mathcal{O}_{L^\infty((0,\ell))}\left(\frac{|f_0|}{\sqrt{\lambda}} + \frac{|f_0'|}{\lambda}\right). \tag{A15}$$

Any component of the eigenfunction φ_n satisfies in the corresponding edge an equation of the form (A12) with some initial data in $x = 0$. Then the discussion on the function $f(x)$ above applies to all the components of the vector φ_n. By the normalization condition $\|\varphi_n\|_{\mathcal{H}^{in}} = 1$ it follows that it must be $\|f\|_{L^2((0,l))} = C$, with $C \leq 1$ (here f denotes a generic component of φ_n, i.e., the restriction of φ_n to a generic edge of $\mathcal{G}^{in}$). Hence, from the identity

$$\int_0^\ell \left| f_0 \cos(\sqrt{\lambda}x) + \frac{f_0'}{\sqrt{\lambda}} \sin(\sqrt{\lambda}x) \right|^2 dx$$
$$= \frac{\ell}{2}\left(|f_0|^2 + \frac{|f_0'|^2}{\lambda}\right) + \frac{\cos(2\sqrt{\lambda}\ell) - 1}{4\sqrt{\lambda}}\left(|f_0|^2 - \frac{|f_0'|^2}{\lambda}\right) + \frac{\mathrm{Re}\,(\bar{f}_0 f_0')}{\lambda}\sin^2(\sqrt{\lambda}\ell)$$

one infers

$$C^2 = \|f\|^2_{L^2((0,l))} = \frac{\ell}{2}\left(|f_0|^2 + \frac{|f_0'|^2}{\lambda}\right) + \mathcal{O}\left(\frac{|f_0|^2}{\sqrt{\lambda}}, \frac{|f_0'|^2}{\lambda^{3/2}}, \frac{|f_0||f_0'|}{\lambda}\right).$$

The latter estimate implies that there exists $\tilde{\lambda}$ such that, for all $\lambda > \tilde{\lambda}$, the inequalities $|f_0| \leq C_1$ and $|f_0'|/\sqrt{\lambda} \leq C_1$ hold true for some positive constant C_1 which does depend on λ. The bounds $|f_0| \leq C_1$ and $|f_0'|/\sqrt{\lambda} \leq C_1$, together with estimate (A15) and the fact that $\lambda_n \to +\infty$ for $n \to \infty$, imply (A11). □

References

1. Kostrykin, V.; Schrader, R. Kirchhoff's rule for quantum wires. *J. Phys. A Math. Gen.* **1999**, *32*, 595–630. [CrossRef]
2. Kostrykin, V.; Schrader, R. Laplacians on metric graphs: eigenvalues, resolvents and semigroups. *Contemp. Math.* **2006**, *415*, 201–226.
3. Berkolaiko, G.; Kuchment, P. Introduction to quantum graphs. In *Mathematical Surveys and Monographs*; American Mathematical Society: Providence, RI, USA, 2013; Volume 186.
4. Berkolaiko, G.; Latushkin, Y.; Sukhtaiev, S. Limits of quantum graph operators with shrinking edges. *arXiv* **2018**, arXiv:1806.00561.
5. Exner, P.; Post, O. Convergence of spectra of graph-like thin manifolds. *J. Geom. Phys.* **2005**, *54*, 77–115. [CrossRef]
6. Exner, P.; Post, O. Quantum networks modelled by graphs. *AIP Conf. Proc.* **2008**, *998*, 1–17.
7. Exner, P.; Post, O. Approximation of quantum graph vertex couplings by scaled Schrödinger operators on thin branched manifolds. *J. Phys. A* **2009**, *42*, 415305. [CrossRef]
8. Post, O. Spectral convergence of quasi-one-dimensional spaces. *Ann. Henri Poincaré* **2006**, *7*, 933–973. [CrossRef]
9. Post, O. *Spectral Analysis on Graph-Like Spaces*; Springer Science & Business Media: Heidelberg, Germany, 2012; Volume 2039.
10. Golovaty, Y.D.; Hryniv, R.O. On norm resolvent convergence of Schrödinger operators with δ'-like potentials. *J. Phys. A Math. Theor.* **2010**, *43*, 155204. [CrossRef]
11. Cacciapuoti, C.; Finco, D. Graph-like models for thin waveguides with Robin boundary conditions. *Asymptot. Anal.* **2010**, *70*, 199–230.
12. Cacciapuoti, C. Graph-like asymptotics for the Dirichlet Laplacian in connected tubular domains. *Anal. Geom. Number Theory* **2017**, *2*, 25–58.
13. Cacciapuoti, C.; Exner, P. Nontrivial edge coupling from a Dirichlet network squeezing: the case of a bent waveguide. *J. Phys. A* **2007**, *40*, F511–F523. [CrossRef]
14. Albeverio, S.; Cacciapuoti, C.; Finco, D. Coupling in the singular limit of thin quantum waveguides. *J. Math. Phys.* **2007**, *48*, 032103. [CrossRef]
15. Exner, P.; Man'ko, S.S. Approximations of quantum-graph vertex couplings by singularly scaled potentials. *J. Phys. A* **2013**, *46*, 345202. [CrossRef]
16. Exner, P.; Man'ko, S.S. Approximations of quantum-graph vertex couplings by singularly scaled rank-one operators. *Lett. Math. Phys.* **2014**, *104*, 1079–1094. [CrossRef]
17. Man'ko, S.S. Schrödinger operators on star graphs with singularly scaled potentials supported near the vertices. *J. Math. Phys.* **2012**, *53*, 123521. [CrossRef]

18. Man'ko, S.S. Quantum-graph vertex couplings: some old and new approximations. *Math. Bohem.* **2014**, *139*, 259–267.
19. Man'ko, S.S. *On δ'-Couplings at Graph Vertices, Mathematical Results in Quantum Mechanics*; World Sci. Publ.: Hackensack, NJ, USA, 2015; pp. 305–313.
20. Cheon, T.; Exner, P.; Turek, O. Approximation of a general singular vertex coupling in quantum graphs. *Ann. Phys.* **2010**, *325*, 548–578. [CrossRef]
21. Ali Mehmeti, F.; Ammari, K.; Nicaise, S. Dispersive effects for the Schrödinger equation on the tadpole graph. *J. Math. Anal. Appl.* **2017**, *448*, 262–280. [CrossRef]
22. Posilicano, A. Self-adjoint extensions of restrictions. *Oper. Matrices* **2008**, *2*, 483–506. [CrossRef]
23. Gohberg, I.; Goldberg, S.; Kaashoek, M.A. *Classes of Linear Operators. Volume I, Operator Theory: Advances and Applications*; Birkhäuser Verlag: Basel, Swiss, 1990.
24. Posilicano, A. A Kreĭn-like formula for singular perturbations of self-adjoint operators and applications. *J. Funct. Anal.* **2001**, *183*, 109–147. [CrossRef]
25. Albeverio, S.; Pankrashkin, K. A remark on Krein's resolvent formula and boundary conditions. *J. Phys. A Math. Gen.* **2005**, *38*, 4859–4864. [CrossRef]
26. Brüning, J.; Geyler, V.; Pankrashkin, K. Spectra of self-adjoint extensions and applications to solvable Schrödinger operators. *Rev. Math. Phys.* **2008**, *20*, 1–70. [CrossRef]
27. Gorbachuk, V. I.; Gorbachuk, M. L. *Boundary Value Problems for Operator Differential Equations, Mathematics and its Applications (Soviet Series)*; Translated and revised from the 1984 Russian original; Kluwer Academic Publishers: Dordrecht, The Netherlands, 1991.
28. Schmüdgen, K. *Unbounded Self-Adjoint Operators on Hilbert Space*, Graduate Texts in Mathematics; Springer: Dordrecht, The Netherlands, 2012; Volume 265.
29. Cacciapuoti, C.; Fermi, D.; Posilicano, A. On inverses of Kreĭn's $\mathscr{Q}$-functions. *Rend. Mat. Appl.* **2018**, *39*, 229–240.
30. Bolte, J.; Endres, S. The trace formula for quantum graphs with general self adjoint boundary conditions. *Ann. Henri Poincaré* **2009**, *10*, 189–223. [CrossRef]
31. Odžak, A.; Šćeta, L. On the Weyl law for quantum graphs. *Bull. Malays. Math. Sci. Soc.* **2019**, *42*, 119–131. [CrossRef]
32. Currie, S.; Watson, B.A. Inverse nodal problems for Sturm-Liouville equations on graphs. *Inverse Probl.* **2007**, *23*, 2029–2040. [CrossRef]
33. Hörmander, L. *Lectures on Nonlinear Hyperbolic Differential Equations*; Mathématiques & Applications (Berlin); Springer: Berlin, Germany, 1997.

symmetry

Article

An Overview on the Standing Waves of Nonlinear Schrödinger and Dirac Equations on Metric Graphs with Localized Nonlinearity

William Borrelli [1,†], Raffaele Carlone [2,†] and Lorenzo Tentarelli [2,*,†]

1 Université Paris-Dauphine, PSL Research University, CNRS, UMR 7534, CEREMADE, F-75016 Paris, France; borrelli@ceremade.dauphine.fr

2 Dipartimento di Matematica e Applicazioni "R. Caccioppoli", Università degli Studi di Napoli "Federico II", MSA, via Cinthia, I-80126 Napoli, Italy; raffaele.carlone@unina.it

* Correspondence: lorenzo.tentarelli@unina.it; Tel.: +39-0816-75696

† These authors contributed equally to this work.

Received: 9 January 2019; Accepted: 27 January 2019; Published: 1 February 2019

Abstract: We present a brief overview of the existence/nonexistence of standing waves for the NonLinear Schrödinger and the NonLinear Dirac Equations (NLSE/NLDE) on metric graphs with localized nonlinearity. First, we focus on the NLSE (both in the subcritical and the critical case) and, then, on the NLDE highlighting similarities and differences with the NLSE. Finally, we show how the two equations are related in the nonrelativistic limit by the convergence of the bound states.

Keywords: metric graphs; NLS; NLD; ground states; bound states; localized nonlinearity; nonrelativistic limit

MSC: 35R02; 35Q55; 81Q35; 35Q40; 49J40; 49J35; 58E05; 46T05

1. Introduction

The aim of this paper is to present the state of the art on the study of the standing waves of the NonLinear Schrödinger and the NonLinear Dirac Equations (NLSE/NLDE) on metric graphs with *localized nonlinearities* (for a summarizing scheme see Section 5).

In the following paper, by *metric graph* we mean the locally compact metric space which one obtains endowing a *multigraph* $\mathcal{G} = (\mathrm{V}, \mathrm{E})$ with a parametrization that associates each bounded edge $e \in \mathrm{E}$ with a closed and bounded interval $I_e = [0, \ell_e]$ of the real line, and each unbounded edge $e \in \mathrm{E}$ with a (copy of the) half-line $I_e = \mathbb{R}^+$ (an extensive description can be found in [1,2] and references therein). Consequently, functions on metric graphs $u = (u_e)_{e\in\mathrm{E}} : \mathcal{G} \to \mathbb{R}, \mathbb{C}$ must be seen as bunches of functions $u_e : I_e \to \mathbb{R}, \mathbb{C}$ such that $u_{|_e} = u_e$. Consistently, Lebesgue and Sobolev spaces are defined as

$$L^p(\mathcal{G}) := \bigoplus_{e\in\mathrm{E}} L^p(I_e), \quad p \in [1,\infty], \qquad \text{and} \qquad H^m(\mathcal{G}) := \bigoplus_{e\in\mathrm{E}} H^m(I_e), \quad m \in \mathbb{N},$$

and are equipped with the natural norms. Moreover, throughout this article, L^p-norms are denoted by $\|u\|_{p,\mathcal{G}}$ and the H^1-norm is denoted by $\|u\|$ for the sake of simplicity.

It is also worth recalling that in the present paper we limit ourselves to focus on the case of metric graphs $\mathcal{G}$ satisfying the following hypothesis:

Hypothesis (H1). *$\mathcal{G}$ is connected and non-compact;*

Hypothesis (H2). *$\mathcal{G}$ has a finite number of edges;*

Hypothesis (H3). *$\mathcal{G}$ has a non-empty compact core $\mathcal{K}$ (which is the subgraph of $\mathcal{G}$ consisting of all its bounded edges);*

(see, e.g., Figure 1).

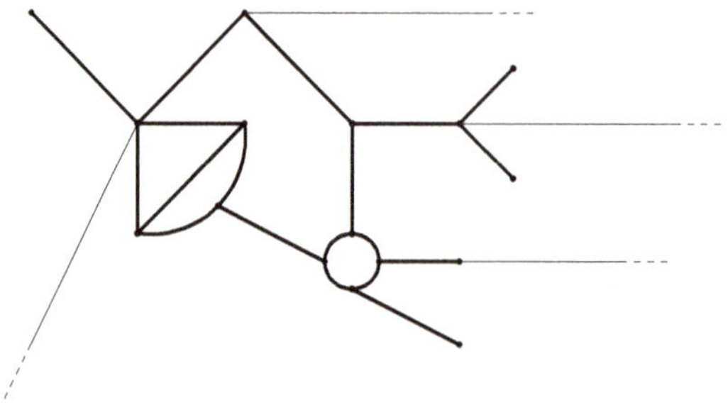

Figure 1. A graph $\mathcal{G}$ satisfying **(H1)**–**(H3)** and its compact core (bold edges).

The growing interest in the study of evolution equations on metric graphs and networks is due to the fact that they are regarded as effective models for the dynamics of systems constrained in branched spatial structures (see [3] and references therein).

In particular, in recent years a considerable attention has been devoted to the *focusing* NLSE, i.e.,

$$\imath \partial_t w = -\Delta_{\mathcal{G}} w - |w|^{p-2}\, w, \qquad p \geq 2, \tag{1}$$

where $-\Delta_{\mathcal{G}}$ is a suitable self-adjoint realization of the operator

$$-\Delta_{|\oplus_{e\in\mathrm{E}} C_0^\infty(\mathring{I}_e)},$$

and, precisely, on the existence of *standing waves* of (1). Those are functions of the form

$$w(t,x) := e^{-i\lambda t}\, u(x),$$

with $\lambda \in \mathbb{R}$ and $u \in L^2(\mathcal{G})$, solving the stationary version of (1), namely,

$$-\Delta_{\mathcal{G}} u - |u|^{p-2}\, u = \lambda u. \tag{2}$$

The physical motivations for the study of the NLSE on metric graphs are extensively explained, e.g., in [3–5] and references therein. We limit ourselves to mention the application to the study of the qualitative behavior of Bose-Einstein condensates in ramified traps (that can be presently realized in laboratory as shown, e.g., by [6]) and to the study of nonlinear optics in Kerr media. In particular, in nonlinear optics one can mention, for instance, the discussion of arrays of planar self-focusing waveguides and of propagation in variously shaped fiber-optic devices (such as Y-junctions, H-junctions, and so on).

The first results in this direction (see, e.g., [7–10]) considered the so-called *infinite N-star* graph (see, e.g., Figure 2) in the case where $-\Delta_{\mathcal{G}}$ is the Laplacian with *δ-type* vertex conditions, that is

$$(-\Delta_{\mathcal{G}}^{\delta,\alpha} u)_{|I_e} := -u_e'', \qquad \forall e \in \mathrm{E}, \qquad \forall u \in \mathrm{dom}(-\Delta_{\mathcal{G}}^{\delta,\alpha}),$$

$$\mathrm{dom}(-\Delta_{\mathcal{G}}^{\delta,\alpha}) := \left\{ u \in H^2(\mathcal{G}) : u \text{ satisfies (3) and (4)} \right\},$$

where

$$u_{e_1}(\mathrm{v}) = u_{e_2}(\mathrm{v}), \qquad \forall e_1, e_2 \succ \mathrm{v}, \qquad \forall \mathrm{v} \in \mathcal{K}, \tag{3}$$

$$\sum_{e \succ v} \frac{du_e}{dx_e}(\mathrm{v}) = \alpha u(\mathrm{v}), \qquad \forall \mathrm{v} \in \mathcal{K}, \tag{4}$$

for some $\alpha \in \mathbb{R}$ ($e \succ \mathrm{v}$ meaning that the edge e is incident at the vertex v, while $\frac{du_e}{dx_e}(\mathrm{v})$ stands for $u_e'(0)$ or $-u_e'(\ell_e)$ according to whether x_e is equal to 0 or ℓ_e at v).

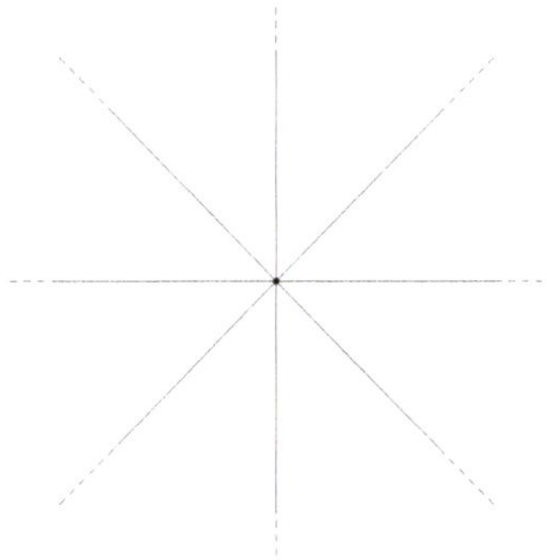

Figure 2. Infinite N-star graph ($N = 8$).

On the other hand, in the case $\alpha = 0$, which is usually called *Kirchhoff* Laplacian (and which will be denoted simply by $-\Delta_{\mathcal{G}}$ in place of $-\Delta_{\mathcal{G}}^{\delta,0}$ in the sequel for the sake of simplicity), more general graphs have been studied (precisely, any graph satisfying (**H1**)–(**H2**)). We mention, in this regard [1,11,12], for a discussion of the existence of *ground states* (i.e., those standing waves that minimize the energy functional associated with (2)) and [13–17] concerning more general *excited states*, a.k.a. *bound states*. We also mention [18] where the same problems are studied in the presence of an external potential.

A modification of this model, introduced in [5,19], consists of assuming that the nonlinearity affects only the compact core of the graph (which then must be supposed non-empty as in (**H3**)) so that (2) reads

$$-\Delta_{\mathcal{G}} u - \chi_{\mathcal{K}} |u|^{p-2} u = \lambda u, \tag{5}$$

where $\chi_{\mathcal{K}}$ is the characteristic function of $\mathcal{K}$. The existence of stationary solutions to (5) has been discussed in [20–22] in the L^2-subcritical case, i.e., $p \in (2, 6)$, and, more recently, in [23,24] in the L^2-critical case, i.e., $p = 6$. A detailed description of these results will be presented in Section 2.

For the sake of completeness, we also remark that the NLSE on compact graphs (which, in particular, do not fulfill (**H1**)) has been studied, e.g., in [25–28]; while the case of one or higher-dimensional *periodic* graphs (which, in particular, do not fulfill (**H2**) as, for instance, in Figure 3) has been addressed, e.g., by [29–32].

Besides the NLS, other dispersive PDEs on metric graphs have been explored in recent years. We mention, for instance, the case of [33] which deals with the dynamics for the Airy equation (motivated by the study of the KdV equation) on star graphs.

On the other hand, other newly studied issues on metric graphs concerns different types of nonlinearities for the NLSE with interesting implications in some physical models (see, e.g., [34]): nonlinearities which are *nonlocal* and/or present a *PT-symmetry* (see, e.g., [35,36]). In particular, the question of the PT-symmetry in quantum graphs has been discussed in the last few years with a focus only on linear problems and, precisely, on the issue of detecting PT-symmetric vertex conditions for symmetric operators on graphs (see [37,38]). Nevertheless, nonlinear PT-symmetric equations on graph are completely unexplored and may represent an interesting topic for future research.

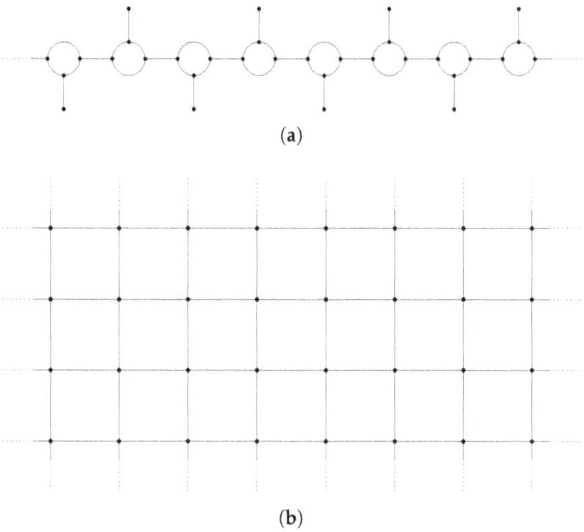

Figure 3. Two examples of periodic graphs: (a) A one-dimensional periodic graph; (b) A two-dimensional grid.

Recently, we started a new research project concerning the NLDE on metric graphs in [39]. The physical motivations for such a model mainly come from solid state physics and nonlinear optics. The existence of Dirac solitons in Bose-Einstein condensates and optical lattices and their concrete realization in discrete waveguide arrays have been investigated in [40,41]. In that case one may expect to recover the metric graph model in an appropriate scaling regime. We also mention that a rigorous mathematical study of the dynamics and the existence of Dirac solitons on lattices has been recently treated in [42–46].

The search for the stationary solutions of the NLDE in this context has been first proposed by [47] for the case of the infinite 3-star graph (see Figure 4). However, in [47] the authors considered the case of an *extended* nonlinearity, i.e.,

$$\imath\partial_t\Psi = \mathcal{D}_\mathcal{G}\Psi - |\Psi|^{p-2}\Psi, \qquad p \geq 2,$$

where $\mathcal{D}_\mathcal{G}$ is a suitable self-adjoint realization of the one-dimensional Dirac operator $\mathcal{D}$, i.e.,

$$\mathcal{D} := -\imath c\frac{d}{dx}\otimes\sigma_1 + mc^2\otimes\sigma_3,$$

$m > 0$ and $c > 0$ representing the *mass* of the generic particle of the system and a *relativistic parameter* (respectively), and σ_1 and σ_3 representing the *Pauli matrices*, i.e.,

$$\sigma_1 := \begin{pmatrix} 0 & 1 \\ 1 & 0 \end{pmatrix} \quad \text{and} \quad \sigma_3 := \begin{pmatrix} 1 & 0 \\ 0 & -1 \end{pmatrix}. \tag{6}$$

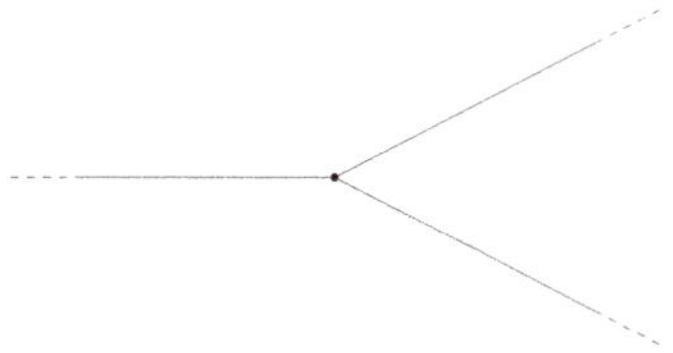

Figure 4. Infinite 3-star graph.

Here the equation of the standing waves $\Psi(t,x) = e^{-\imath\omega t}\psi(x)$ reads

$$\mathcal{D}_{\mathcal{G}}\psi - |\psi|^{p-2}\,\psi = \omega\psi.$$

In fact, in order to deal with more complex graph topologies, in [39] we restricted ourselves to the case of a Kirchhoff-type extension of the Dirac operator (for details see Section 3.1) and, most importantly, we considered the case of a *localized* nonlinearity, that is

$$\mathcal{D}_{\mathcal{G}}\psi - \chi_{\mathcal{K}}|\psi|^{p-2}\,\psi = \omega\psi. \tag{7}$$

In addition, in order to test the consistency of this model we also proved (see Section 4) the convergence (of the bound states) of (7) to (the bound states of) (5) in the *nonrelativistic limit*, namely as "$c \to \infty$".

Please note that in (7) (as in the equation studied in [47]) the nonlinearity is a pure power, thus being manifestly *non-covariant*. This kind of nonlinearities typically arises in nonlinear optics. We also stress that from a theoretical point of view there is no conceptual contradiction as the NLDE should be intended only as an effective model.

Nevertheless, the equation obtained replacing the nonlinearity $|\psi|^{p-2}\,\psi$ with a *covariant* one, i.e.,

$$\mathcal{D}_{\mathcal{G}}\psi - |\langle\psi,\sigma_3\psi\rangle|^{(p-2)/2}\psi = \omega\psi$$

(the nonlinearity now being defined in terms of a Lorentz-scalar) is of great interest. This equation, indeed, might be interpreted as the model for a *truly relativistic* particle (a Dirac fermion), whose dynamics is constrained to a one-dimensional structure. In this case, the model should be Lorentz-covariant (strictly speaking, in this case the equation should be rewritten in covariant form). We expect that this will require some technical adjustments to deal with the indefiniteness of the nonlinearity. This might be quite relevant in the study of the nonrelativistic limit, since in [39] we used the fact that a positive nonlinearity allows to get a priori estimates uniform in c.

Finally, we mention that the discussion of extended nonlinearities presents some technical problems that we were not able to overcome thus far. However, we plan to investigate this question in a forthcoming paper.

2. Nonlinear Schrödinger Equation

The study of the existence/nonexistence and multiplicity of bound states of (5) carried on in [20–24] is completely based on variational methods.

This is due to the simple observation that L^2-solutions of (5), with $-\Delta_{\mathcal{G}}$ denoting the Kirchhoff Laplacian, arise also as *constrained* critical points of the energy functional

$$E(u,\mathcal{K},p) := \frac{1}{2}\int_{\mathcal{G}}|u'|^2\,dx - \frac{1}{p}\int_{\mathcal{K}}|u|^p\,dx, \qquad p \geq 2,$$

on the manifold

$$\mathcal{H}_\mu(\mathcal{G}) := \left\{u \in \mathcal{H}(\mathcal{G}) : \|u\|^2_{2,\mathcal{G}} = \mu\right\} \tag{8}$$

for some fixed (but arbitrary) $\mu > 0$, where

$$\mathcal{H}(\mathcal{G}) := \left\{ u \in H^1(\mathcal{G}) : u \text{ satisfies } (3) \right\}.$$

Moreover, if u is a constrained critical point, also known as *bound state*, then computation shows that the Lagrange multiplier λ can be found as

$$\lambda = \lambda(u) := \frac{1}{\mu}\left(\int_{\mathcal{G}} |u'|^2\,dx - \int_{\mathcal{K}} |u|^p\,dx \right). \tag{9}$$

Remark 1. *Please note that $\mathcal{H}(\mathcal{G})$ is the form domain of $-\Delta_{\mathcal{G}}$. In addition, it is worth mentioning that the choice of the manifold (8) is also related to the fact that the time dependent counterpart of (5) is mass-preserving.*

2.1. Ground States

The first step in the study of the bound states consists in looking for *ground states*, namely the constrained minimizers of $E(\cdot, \mathcal{K}, p)$ on the manifold $\mathcal{H}_\mu(\mathcal{G})$.

Ground states can be shown (see [1]) to be of the form

$$u(x) = e^{i\phi} v(x)$$

where ϕ is a fixed phase factor and v is real-valued. Hence, in the following, in minimization problems we will always limit ourselves to consider real-valued functions.

For more general bound states we cannot prove at the moment that such a property holds. Therefore, while in existence and multiplicity results we will limit to real-valued functions, nonexistence results concern also complex-valued functions.

2.1.1. The Subcritical Case: $p \in (2, 6)$

We start reporting on the existence of constrained minimizers in the so-called L^2-subcritical case, namely, when $p \in (2, 6)$.

For $p > 2$ and for every graph satisfying **(H1)**–**(H2)** it is possible to prove (see, e.g., [11], [Proposition 2.1] and [22], [Proposition 4.1] for real-valued functions and [21], [Proposition 2.6] for complex-valued functions) that the following *Gagliardo-Nirenberg inequalities* hold:

$$\|u\|_{p,\mathcal{G}}^p \le C(\mathcal{G}, p) \|u'\|_{2,\mathcal{G}}^{\frac{p}{2}-1} \|u\|_{2,\mathcal{G}}^{\frac{p}{2}+1}, \qquad \forall u \in \mathcal{H}(\mathcal{G}), \tag{10}$$

$$\|u\|_{\infty,\mathcal{G}} \le C(\mathcal{G}, \infty) \|u'\|_{2,\mathcal{G}}^{\frac{1}{2}} \|u\|_{2,\mathcal{G}}^{\frac{1}{2}}, \qquad \forall u \in \mathcal{H}(\mathcal{G}) \tag{11}$$

with $C(\mathcal{G}, p)$ and $C(\mathcal{G}, \infty)$ denoting the optimal constants of the respective inequalities. Then

$$E(u, \mathcal{K}, p) \ge \frac{1}{2}\|u'\|_{2,\mathcal{G}}^2 - \frac{C(\mathcal{G}, p)}{p} \|u'\|_{2,\mathcal{G}}^{\frac{p}{2}-1} \|u\|_{2,\mathcal{G}}^{\frac{p}{2}+1}, \qquad \forall u \in \mathcal{H}(\mathcal{G}),$$

so that

$$E(u, \mathcal{K}, p) \ge \frac{1}{2}\|u'\|_{2,\mathcal{G}}^{\frac{p}{2}-1} \left(\|u'\|_{2,\mathcal{G}}^{\frac{6-p}{2}} - \frac{2\mu^{\frac{p+2}{4}} C(\mathcal{G}, p)}{p} \right), \qquad \forall u \in \mathcal{H}_\mu(\mathcal{G}), \tag{12}$$

and, if $p \in (2, 6)$, this immediately entails that the functional $E(\cdot, \mathcal{K}, p)$ is bounded from below on the manifold $\mathcal{H}_\mu(\mathcal{G})$.

Remark 2. *Please note that in fact, (10) and (11) hold for every $u \in H^1(\mathcal{G})$ under the sole assumption* **(H2)**, *up to a redefinition of the optimal constants. However, we chose to mention the $\mathcal{H}(\mathcal{G})$-version, which holds even* *if* (**H2**) *is not fulfilled. We remark that the best constants of these versions play a crucial role in the analysis of the ground states.*

In addition to (12), as a general fact, one can prove that

$$\inf_{u\in\mathcal{H}_\mu(\mathcal{G})} E(u,\mathcal{K},p) \leq 0 \tag{13}$$

and that

$$\inf_{u\in\mathcal{H}_\mu(\mathcal{G})} E(u,\mathcal{K},p) < 0 \qquad \Longrightarrow \qquad \text{a ground state does exist.} \tag{14}$$

As a consequence, the following results can be established

Theorem 1 ([22] [Theorems 3.3 and 3.4], [21] [Corollary 3.4])**.** *Let $\mathcal{G}$ satisfy* (**H1**)–(**H3**) *and let $p \in (2,6)$. Therefore:*

(1) *if $p < 4$, then there exists a ground state for every $\mu > 0$;*
(2) *if $p \geq 4$, then:*

(i) *whenever*

$$\mu^{\frac{p-2}{6-p}}|\mathcal{K}| > N^{\frac{4}{6-p}}c_p, \tag{15}$$

where N is the number of half-lines of $\mathcal{G}$ and

$$c_p := \left[\left(\frac{p(p-4)}{16}\right)^{\frac{2}{p-2}} + \frac{p}{8}\left(\frac{p(p-4)}{16}\right)^{\frac{4-p}{p-2}}\right]^{\frac{p-2}{6-p}},$$

there exists a ground state of mass μ;

(ii) *whenever*

$$\mu^{\frac{p-2}{6-p}}|\mathcal{K}| < \left(\frac{p}{2}\right)^{\frac{2}{6-p}}\frac{C(\mathcal{G},p)^{\frac{4-p}{6-p}}}{C(\mathcal{G},\infty)^p}, \tag{16}$$

there does not exist any ground state of mass μ.

The proof of the previous theorem is clearly based on (13) and (14). Precisely, from (14) there results that the existence part can be obtained exhibiting a function with negative energy, usually called *competitor*. This can be achieved for every mass μ when $p \in (2,4)$, while when $p \in [4,6)$ this is possible only under (15). Such competitors are constant on the compact core and exponentially decreasing on half-lines. In this way they minimize the kinetic energy on the compact core and have the exact qualitative behavior of the bound states on the half-lines (where the problem is linear). Clearly, such competitors cannot be minimizers since they do not fulfill (4) with $\alpha = 0$.

On the other hand, (13) shows that to prove nonexistence, it is sufficient to prove that every function possesses a strictly positive energy level. This is, actually, the idea behind the first proof of the nonexistence result for ground states. However, there is also another strategy, which is similar to that of the proof of Theorem 5, which allows to prove item 2)(ii) in a straighter way and with the sharper threshold given by (16).

Finally, it is worth mentioning both that the negativity of the energy levels of the minimizers entails that the associated Lagrange multipliers (given by (9)) are negative, and that the results of Theorem 1 are invariant under the *homotetic* transformation

$$\mu \mapsto \sigma\mu, \qquad \mathcal{G} \mapsto \sigma^{\frac{2-p}{6-p}}\mathcal{G}.$$

In other words, if one sets for instance $p = 4$, then a problem on a graph $\mathcal{G}$ with a given mass constraint μ is completely equivalent to a problem with half the mass and a "doubled" graph, and vice versa.

2.1.2. The Critical Case: $p = 6$

The study of the so-called L^2-critical case $p = 6$ has been successfully managed only very recently. Here the main difficulty comes from the fact that the boundedness from below of the energy functional strongly depends on the mass constraint.

Indeed, in this case (12) reads

$$E(u,\mathcal{K},6) \geq \frac{1}{2}\|u'\|_{2,\mathcal{G}}^2 \left(1 - \frac{\mu^2 C(\mathcal{G},6)}{3}\right), \qquad \forall u \in \mathcal{H}_\mu(\mathcal{G}), \tag{17}$$

and hence the boundedness of functional clearly depends on the value of μ. In particular, it turns out that the discriminating value is related to the best constant of (10); that is, the *critical mass* $\mu_{\mathcal{G}}$ of the problem is defined by

$$\mu_{\mathcal{G}} := \sqrt{\frac{3}{C(\mathcal{G},6)}}.$$

It is well known that when $\mathcal{G} = \mathbb{R}, \mathbb{R}^+$

$$\mu_{\mathbb{R}} = \frac{\pi\sqrt{3}}{2}, \qquad \mu_{\mathbb{R}}^+ = \frac{\mu_{\mathbb{R}}}{2}, \tag{18}$$

that

$$\mu_{\mathbb{R}}^+ \leq \mu_{\mathcal{G}} \leq \mu_{\mathbb{R}}, \qquad \forall \mathcal{G} \text{ satisfying } (\mathbf{H1})\text{–}(\mathbf{H3}),$$

and that $\mu_{\mathbb{R}}, \mu_{\mathbb{R}}^+$ are the sole values of the mass constraint at which the problem with the extended nonlinearity on $\mathbb{R}, \mathbb{R}^+$ (respectively) admits a ground state (see, e.g., [48]).

However, one can easily see that, while for the problem with extended nonlinearity (treated in [12]) $\mu_{\mathcal{G}}$ is the proper parameter to be investigated, in the localized case one should better consider a *reduced* critical mass

$$\mu_{\mathcal{K}} := \sqrt{\frac{3}{C(\mathcal{K})}}, \tag{19}$$

where $C(\mathcal{K})$ denotes the optimal constant of the following modified Gagliardo-Nirenberg inequality

$$\|u\|_{6,\mathcal{K}}^6 \leq C(\mathcal{K})\|u'\|_{2,\mathcal{G}}^2 \|u\|_{2,\mathcal{G}}^4, \qquad \forall u \in \mathcal{H}(\mathcal{G}),$$

namely

$$C(\mathcal{K}) := \sup_{u \in \mathcal{H}(\mathcal{G})} \frac{\|u\|_{6,\mathcal{K}}^6}{\|u'\|_{2,\mathcal{G}}^2 \|u\|_{2,\mathcal{G}}^4}.$$

In fact, using this new parameter one obtains

$$E(u,\mathcal{K},6) \geq \frac{1}{2}\|u'\|_{2,\mathcal{G}}^2 \left(1 - \frac{\mu^2 C(\mathcal{K})}{3}\right), \qquad \forall u \in \mathcal{H}_\mu(\mathcal{G}),$$

which is clearly sharper than (17), as one can easily see that $C(\mathcal{K}) \leq C(\mathcal{G},6)$.

Existence of ground states has been first investigated in [23] in the simple case of the *tadpole* graph (see Figure 5). Very recently more refined results have been obtained without any assumption on the topology of the graph, thus providing a (almost) complete classification of the phenomenology.

Figure 5. Tadpole graph.

To state such results, let

$$\mathcal{E}_{\mathcal{G}}(\mu) := \inf_{u \in \mathcal{H}_\mu(\mathcal{G})} E(u, \mathcal{K}, 6),$$

and recall the following definitions for a metric graph:

- a graph $\mathcal{G}$ is said to admit a *cycle-covering* if and only if every edge of $\mathcal{G}$ belongs to a *cycle*, namely either a *loop* (i.e., a closed path of consecutive bounded edges) or an unbounded path joining the endpoints of two distinct half-lines (which are then identified as a single vertex at infinity);
- a graph $\mathcal{G}$ is said to possesses a *terminal edge* if and only if it contains an edge ending with a vertex of degree one.

Theorem 2 ([24] [Theorem 1.1]). *Let $\mathcal{G}$ satisfy* **(H1)**–**(H3)**. *Then,*

$$\mathcal{E}_{\mathcal{G}}(\mu, \mathcal{K}) \begin{cases} = 0 & \text{if } \mu \leq \mu_{\mathcal{K}} \\ < 0 & \text{if } \mu \in (\mu_{\mathcal{K}}, \mu_{\mathbb{R}}] \\ = -\infty & \text{if } \mu > \mu_{\mathbb{R}}. \end{cases}$$

In addition,

(i) *if $\mathcal{G}$ has at least a terminal edge (as, for instance, in Figure 6a), then*

$$\mu_{\mathcal{K}} = \mu_{\mathbb{R}^+}, \qquad \mathcal{E}_{\mathcal{G}}(\mu, \mathcal{K}) = -\infty \quad \text{for all } \mu > \mu_{\mathcal{K}},$$

and there is no ground state of mass μ for any $\mu > 0$;

(ii) *if $\mathcal{G}$ admits a cycle-covering (as, for instance, in Figure 6b), then*

$$\mu_{\mathcal{K}} = \mu_{\mathbb{R}}$$

and there is no ground state of mass μ for any $\mu > 0$;

(iii) *if $\mathcal{G}$ has only one half-line and no terminal edges (as, for instance, in Figure 6c), then*

$$\mu_{\mathbb{R}^+} < \mu_{\mathcal{K}} < \sqrt{3} \tag{20}$$

and there is a ground states of mass μ if and only if $\mu \in [\mu_{\mathcal{K}}, \mu_{\mathbb{R}}]$.

(iv) *if $\mathcal{G}$ has no terminal edges, does not admit a cycle-covering, but presents at least two half-lines (as, for instance, in Figure 6d), then*

$$\mu_{\mathbb{R}^+} < \mu_{\mathcal{K}} \leq \mu_{\mathbb{R}} \tag{21}$$

and there is a ground states of mass μ if and only if $\mu \in [\mu_{\mathcal{K}}, \mu_{\mathbb{R}}]$, provided that $\mu_{\mathcal{K}} \neq \mu_{\mathbb{R}}$.

Theorem 2 presents several differences with respect to its analogous for the extended case established in [12] (in addition to the fact that $\mu_{\mathcal{G}}$ is replaced with $\mu_{\mathcal{K}}$). For details on such differences we refer the reader to [24]. Here, we aim at highlighting a more crucial feature of the previous result. Indeed, while in the subcritical case the topology of the graph plays no role in establishing existence/nonexistence of ground states, now the topological classification of the graphs is the central point as well as it occurs in the context of extended nonlinearities. Even though there is no rigorous explanation of such a phenomenon, this seems to be related to the fact that the critical power of the nonlinearity "breaks" the characteristic length of the graph, i.e., $\mu^{\frac{p-2}{6-p}}|\mathcal{K}|$, thus making the metric properties less relevant in the discussion.

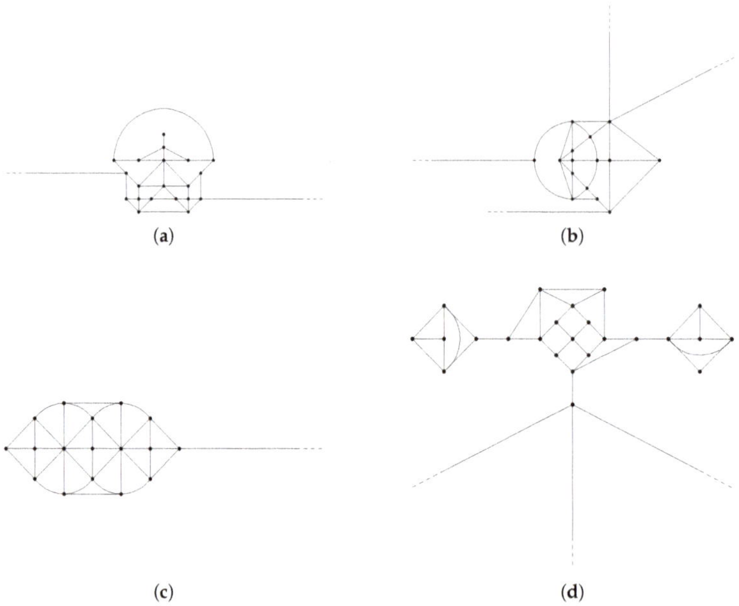

Figure 6. Examples of graphs from cases *(i)-(iv)* of Theorem 2: (**a**) A graph with one terminal edge; (**b**) A graph admitting a cycle-covering; (**c**) A graph with one half-line and without terminal edges; (**d**) A graph without terminal edges and cycle-coverings, and with two half-lines.

Nevertheless, they preserve a role also in the critical case, at least in cases (iii) and (iv), as shown by the following

Theorem 3 ([24] [Theorems 1.2 and 1.3])**.** *Estimates* (20) *and* (21) *are sharp in general; i.e., for every* $\varepsilon > 0$ *there exist two non-compact metric graphs* $\mathcal{G}_\varepsilon^1, \mathcal{G}_\varepsilon^2$ *(with compact cores* $\mathcal{K}_\varepsilon^1, \mathcal{K}_\varepsilon^2$*) with one half-line and without terminal edges such that*

$$\mu_{\mathcal{K}_\varepsilon^1} \leq \mu_{\mathbb{R}^+} + \varepsilon \qquad \text{and} \qquad \mu_{\mathcal{K}_\varepsilon^2} \geq \sqrt{3} - \varepsilon,$$

and two non-compact metric graphs $\mathcal{G}_\varepsilon^3, \mathcal{G}_\varepsilon^4$ *(with compact cores* $\mathcal{K}_\varepsilon^3, \mathcal{K}_\varepsilon^4$*) without terminal edges and cycle-coverings and with at least two half-lines such that*

$$\mu_{\mathcal{K}_\varepsilon^3} \leq \mu_{\mathbb{R}^+} + \varepsilon \qquad \text{and} \qquad \mu_{\mathcal{K}_\varepsilon^4} \geq \mu_{\mathbb{R}} - \varepsilon.$$

In other words, Theorem 3 shows that estimates (20) and (21) are sharp by exhibiting four suitable sequences of graphs. They can be constructed as follows:

(1) the sequence $\mathcal{G}_\varepsilon^1$ can be constructed by considering a graph whose compact core does not admit a cycle-covering (see, e.g., Figure 7a) and letting the length of one of its *cut-edges*, the edges whose removal disconnects the graph (e.g., $\widehat{e}$ in Figure 7a), go to infinity;

(2) the sequence $\mathcal{G}_\varepsilon^2$ can be constructed by considering a graph as in Figure 7b and letting the length of the compact core go to infinity keeping at the same time the total diameter of the compact core bounded (namely, thickening the compact core);
(3) the sequences $\mathcal{G}_\varepsilon^3, \mathcal{G}_\varepsilon^4$ can be constructed by considering a *signpost* graph (see, e.g., Figure 7c) and letting the length of its cut-edge $\tilde{e}$ go to infinity and to zero, respectively.

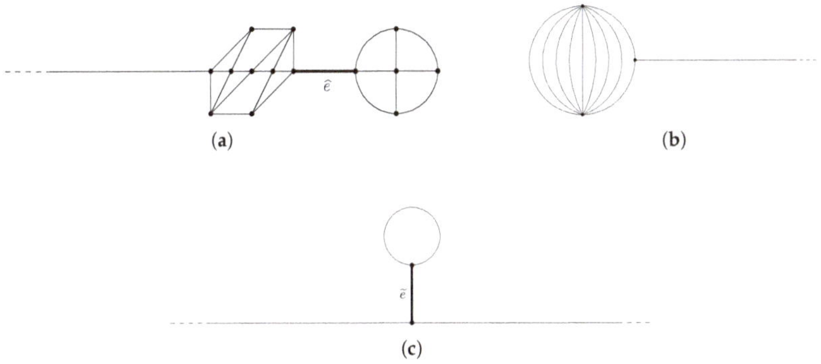

Figure 7. Examples of graphs from Theorem 3: (**a**) A graph whose compact core does not admit a cycle-covering; (**b**) A tadpole graph with extra connections between two vertices; (**c**) A signpost graph.

The proofs of Theorems 2 and 3 are quite technical and we refer the reader to [24] for a complete presentation. They are based on a careful analysis of the behavior of the constant $C(\mathcal{K})$ and on the "graph surgery" and rearrangement techniques developed by [1,11,12]. In particular, given a nonnegative function $u \in \mathcal{H}(\mathcal{G})$, one uses the *decreasing* rearrangement

$$u^*(x) := \inf\{t \geq 0 : \rho(t) \leq x\}, \quad x \in [0, |\mathcal{G}|)$$

and the *symmetric* rearrangement

$$\widehat{u}(x) := \inf\{t \geq 0 : \rho(t) \leq 2|x|\}, \qquad x \in (-|\mathcal{G}|/2, |\mathcal{G}|/2),$$

where

$$\rho(t) := \sum_{e \in \mathrm{E}} |\{x_e \in I_e : u_e(x_e) > t\}| \quad t \geq 0,$$

and precisely the facts that

$$\|u\|_{p,\mathcal{G}} = \|u^*\|_{p,\mathbb{R}^+} = \|\widehat{u}\|_{p,\mathbb{R}}, \qquad \forall p \geq 1,$$

that

$$\|(u^*)'\|_{2,\mathbb{R}^+} \leq \|u'\|_{2,\mathcal{G}}$$

and that for every function u which possesses at least two preimages at each level,

$$\|(\widehat{u})'\|_{2,\mathbb{R}} \leq \|u'\|_{2,\mathcal{G}},$$

to construct suitable minimizing sequences or suitable competitors to exploit level arguments analogous to (14) to get compactness in the limit.

2.2. Bound States

Besides the investigation of ground states, one can also study standing waves which are not necessarily minimizers of the constrained energy, namely *bound/excited states*. They may arise as constrained critical points of the energy functional on $\mathcal{H}_\mu(\mathcal{G})$.

To the best of our knowledge all the known results on existence (and multiplicity), as well as on nonexistence, of general bound states strongly exploit the assumption on the subcriticality of the power nonlinearity (up to some exception that we mention below).

2.2.1. Existence Results

Existence and multiplicity results in this context have been proved by extending some well-known techniques from Critical Point Theory (see, e.g., [49–51]).

However, the context of metric graphs presents two additional technical issues. First, it is not possible, in general, to gain compactness restricting to symmetric (in some suitable sense) functions if one does not want to restrict the discussion to symmetric graphs. Furthermore, the fact that graphs with non-empty compact core are not homothetically invariant entails that multiple solutions of prescribed mass cannot be found by scaling arguments.

The strategy used in [20], to solve these problems is, then, the following:

- detecting the energy levels at which the *Palais-Smale condition* is satisfied, namely detect the values $c \in \mathbb{R}$ such that any sequence $(u_n) \in \mathcal{H}_\mu(\mathcal{G})$ satisfying

  (i) $E(u_n, \mathcal{K}, p) \to c$
  (ii) $\|dE_{|\mathcal{H}_\mu(\mathcal{G})}(u_n, \mathcal{K}, p)\|_{\mathbb{T}'_{u_n}\mathcal{H}_\mu(\mathcal{G})} \to 0$

  (with $\mathbb{T}'_{u_n}\mathcal{H}_\mu(\mathcal{G})$ denoting the topological dual of the tangent to the manifold $\mathcal{H}_\mu(\mathcal{G})$ at u_n) admits a subsequence converging in $\mathcal{H}_\mu(\mathcal{G})$;
- constructing suitable *min-max* levels.

In this case, one can prove that the constrained functional possesses the Palais-Smale property only at negative levels and whence the min-max levels must be negative. In addition, as the functional is even, it is possible to construct such min-max levels using the *Krasnosel'skii genus* (see, e.g., [52]), that is, for every $A \subset \mathcal{H}(\mathcal{G})\backslash\{0\}$ closed and symmetric, the natural number defined by

$$\gamma(A) := \min\{n \in \mathbb{N} : \exists\phi : A \to \mathbb{R}^n\backslash\{0\},\ \phi \text{ continuous and odd}\}.$$

In fact, one can prove that the suitable levels are

$$c_j := \inf_{A \in \Gamma_j} \max_{u \in A} E(u, \mathcal{K}, p),$$

where

$$\Gamma_j := \{A \subset \mathcal{H}_\mu(\mathcal{G}) : A \text{ is compact and symmetric, and } \gamma(A) \geq j\}.$$

In this way it is possible to prove the following

Theorem 4 ([20] [Theorem 1.2]). *Let $\mathcal{G}$ satisfy* **(H1)**–**(H3)** *and let $p \in (2,6)$. For every $k \in \mathbb{N}$, there exists $\mu_k > 0$ such that for every $\mu \geq \mu_k$ there exist at least k distinct pairs $(\pm u_j)$ of bound states of mass μ. Moreover, for every $j = 1, \ldots, k$,*

$$E(u_j, \mathcal{K}, p) \leq jE(\varphi_{\mu/j}, \mathbb{R}, p) + \sigma_k(\mu) < 0,$$

where $\varphi_{\mu/j}$ denote the unique positive minimizer of $E(\cdot, \mathbb{R}, p)$ constrained on $\mathcal{H}_{\mu/j}(\mathcal{G})$ and $\sigma_k(\mu) \to 0$ (exponentially fast) as $\mu \to \infty$. Finally, for each j, the Lagrange multiplier λ_j relative to u_j is negative.

The strategy hinted before cannot be easily adapted to the extended problem since in this case the Palais-Smale condition fails also at infinitely many negative energy levels and then it is open how to fix a min-max level in the gap between two of these levels.

On the other hand, it is also unclear how to extend this strategy to the critical case. Indeed, the construction of negative energy min-max levels is based on the possibility of putting (on increasing the mass) several suitable truncations of scaled copies of $\varphi_{\mu/j}$ on the compact core of the graph keeping the total energy negative. However, no direct analogous of such a technique is available for the critical case as the minimizers φ_μ do exist only at the critical mass and present a zero energy level.

2.2.2. Nonexistence Results

Nonexistence results concern clearly only the regime $p \in [4,6)$ since for $p \in (2,4)$ the existence of bound states is guaranteed by Theorem 1.

The interesting part of nonexistence results for bound states in the subcritical regime is that they strongly rely both on metric and on topological features of the graph as one can see by the following

Theorem 5 ([21] [Theorems 3.2 and 3.5]). *Let $\mathcal{G}$ satisfy* **(H1)**–**(H3)** *and let $p \in [4,6)$. Therefore,*

(i) *if the graph $\mathcal{G}$ satisfies*

$$\mu^{\frac{p-2}{6-p}}|\mathcal{K}| < \frac{C(\mathcal{G},p)^{\frac{4-p}{6-p}}}{C(\mathcal{G},\infty)^p},$$

then, there are no bound states of mass μ with $\lambda \le 0$;

(ii) *if $\mathcal{G}$ is a* tree *(i.e., no loops) with at most one pendant (see, e.g., Figure 8), then there is no bound state of mass μ with $\lambda \ge 0$, for every $\mu > 0$.*

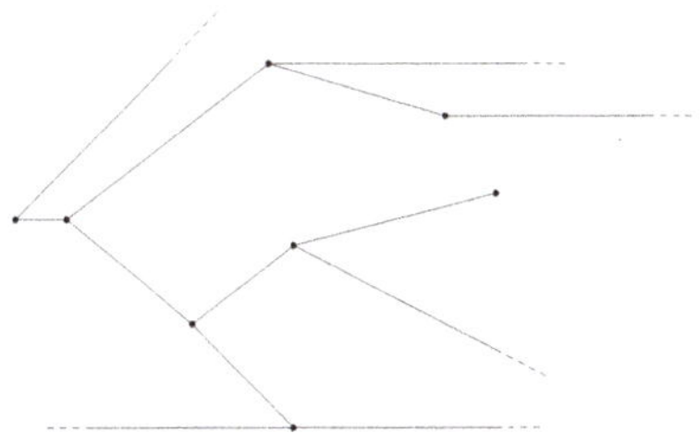

Figure 8. A tree with one pendant.

The most remarkable fact is that the dependence on metric or topological features in the nonexistence result is connected to the sign of the Lagrange multiplier. In particular, the first condition prevents the existence of bound states supported on the whole $\mathcal{G}$, as such functions cannot possess a negative Lagrange multiplier since they are in $L^2(\mathcal{G})$. On the other hand, one can check that the second part of Theorem 5 holds as well for $p \ge 6$ and even in the extended nonlinearity case.

Finally, it is worth observing that the condition of the second part of Theorem 5 is sharp. Indeed, one can easily construct counterexamples whenever the assumption of being a tree with at most one pendant is dropped. Precisely, if $\mathcal{G}$ possesses a loop, then one can define a bound state with a nonnegative Lagrange multiplier supported only on the loop (see, e.g., Figure 9a); if $\mathcal{G}$ is a tree with two pendants, instead, then one can define a bound state with nonnegative Lagrange multiplier supported on the path which joins the two pendants (see, e.g., Figure 9b).

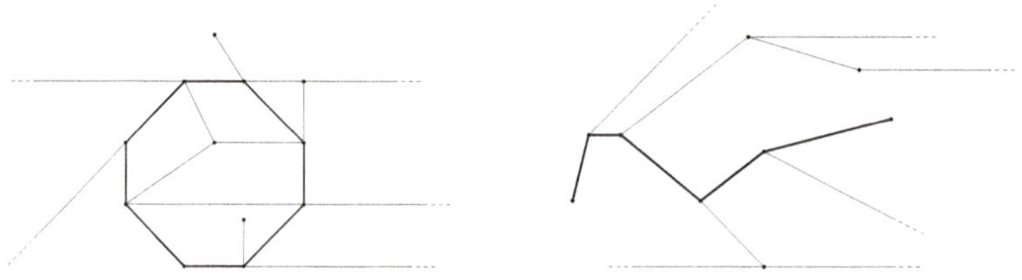

(**a**) (**b**)

Figure 9. Examples of graphs for which item (ii) of Theorem 5 does not hold: (**a**) A graph with a loop. (**b**) A tree with two pendants.

3. Nonlinear Dirac Equation

As already mentioned in Section 1, the study of bound state-solutions of NLDE (7) in the context of metric graphs has been introduced in [47]. To our knowledge, the first rigorous work on this subject for general graphs has been done in [39] for the case of localized nonlinearities.

Such discussion requires a suitable definition of the Dirac operator on graphs and the adaptation of some techniques from Critical Point Theory.

Before going more into details, we recall that Dirac-type equations are of spinorial nature and, hence, here we de deal with vector-valued function on graphs (2-*spinors*). More precisely, we consider wave functions $\psi : \mathcal{G} \to \mathbb{C}^2$ that can be seen as a family of 2-spinors on intervals, i.e., $\psi = (\psi_e)$ with

$$\psi_e = \begin{pmatrix} \phi_e \\ \chi_e \end{pmatrix} : I_e \longrightarrow \mathbb{C}^2, \qquad \forall e \in \mathrm{E},$$

or equivalently as 2-components vector of functions on graphs, i.e., $\psi = (\phi, \chi)^T$, with $\phi = (\phi_e)$ and $\chi = (\chi_e)$. Accordingly,

$$L^p(\mathcal{G}, \mathbb{C}^2) := \bigoplus_{e \in \mathrm{E}} L^p(I_e, \mathbb{C}^2) = \{\psi : \mathcal{G} \to \mathbb{C}^2 : \phi, \chi \in L^p(\mathcal{G})\},$$

and

$$H^m(\mathcal{G}, \mathbb{C}^2) := \bigoplus_{e \in \mathrm{E}} H^m(I_e, \mathbb{C}^2) = \{\psi : \mathcal{G} \to \mathbb{C}^2 : \phi, \chi \in H^m(\mathcal{G})\},$$

endowed with the natural norms, which we denote, with a little abuse of notation, by $\|\psi\|_{p,\mathcal{G}}$ and $\|\psi\|$ (in the case $m = 1$), respectively.

3.1. Remarks on the Dirac Operator on Graphs

As we mentioned before, the first step in the study of the NLDE on graphs is to define a self-adjoint realization of the Dirac operator.

A complete discussion of such a topic can be found, e.g., in [53,54]. Here, however, we limit ourselves to the extension that we call of *Kirchhoff type*, since it represents the analogous to the Schrödinger operator with Kirchhoff conditions and, hence, corresponds to the *free case*, that is, the case where there is no interaction at vertices (for detail on such extension, refer to [39]).

For every fixed m, $c > 0$, we define the Dirac operator of Kirchhoff type as the operator

$$\mathcal{D}_{\mathcal{G}} : L^2(\mathcal{G}, \mathbb{C}^2) \to L^2(\mathcal{G}, \mathbb{C}^2)$$

with action

$$\mathcal{D}_{\mathcal{G}|I_e}\psi = \mathcal{D}_e\psi_e := -\imath c\,\sigma_1\psi_e' + mc^2\sigma_3\psi_e, \qquad \forall e \in \mathrm{E}, \qquad \forall \psi \in \mathrm{dom}(\mathcal{D}_{\mathcal{G}}),$$

σ_1, σ_3 being the Pauli matrices defined in (6), and whose domain is

$$\mathrm{dom}(\mathcal{D}_\mathcal{G}) := \left\{ \psi \in H^1(\mathcal{G}, \mathbb{C}^2) : \psi \text{ satisfies (22) and (23)} \right\},$$

where

$$\phi_{e_1}(\mathrm{v}) = \phi_{e_2}(\mathrm{v}), \qquad \forall e_1, e_2 \succ \mathrm{v}, \qquad \forall \mathrm{v} \in \mathcal{K}, \tag{22}$$

$$\sum_{e \succ \mathrm{v}} \chi_e(\mathrm{v})_\pm = 0, \qquad \forall \mathrm{v} \in \mathcal{K}, \tag{23}$$

$\chi_e(\mathrm{v})_\pm$ standing for $\chi_e(0)$ or $-\chi_e(\ell_e)$ according to whether x_e is equal to 0 or ℓ_e at v.

Such operator can be proved to be self-adjoint and, in addition, possesses spectral properties analogous to the standard Dirac operator on $\mathbb{R}$, that is

$$\sigma(\mathcal{D}_\mathcal{G}) = (-\infty, -mc^2] \cup [mc^2, +\infty), \tag{24}$$

even though, according to the structure of the graph, there might be eigenvalues embedded at any point of the spectrum.

The actual reason for which we call such operator of Kirchhoff type is better explained by Section 4. However, an informal justification is provided by the following formal computation. First, one can see that $\mathcal{D}_\mathcal{G}^2$ acts as $(-\Delta_\mathcal{G}) \otimes \mathbb{I}_{\mathbb{C}^2}$ (if $m = 0$). In addition, if one considers spinors of the type $\psi = (\phi, 0)^T$ and assumes that they belong to the domain of $\mathcal{D}_\mathcal{G}^2$, namely that $\psi \in \mathrm{dom}(\mathcal{D}_\mathcal{G})$ and that $\mathcal{D}_\mathcal{G}\psi \in \mathrm{dom}(\mathcal{D}_\mathcal{G})$, one clearly sees that $\phi \in \mathrm{dom}(-\Delta_\mathcal{G})$.

Finally, compared to the Schrödinger case, a big difference is given by the definition of the *quadratic form* associated with $\mathcal{D}_\mathcal{G}$. In particular, here it is not explicitly known since Fourier transform is not available (in a simple way), as we are not in an Euclidean space, and since classical duality arguments fail, as it is not true in general that $H^{-1/2}(\mathcal{G}, \mathbb{C}^2)$ is the topological dual of $H^{1/2}(\mathcal{G}, \mathbb{C}^2)$.

It can be defined, clearly, using the *Spectral Theorem*, so that

$$\mathrm{dom}(\mathcal{Q}_{\mathcal{D}_\mathcal{G}}) := \left\{ \psi \in L^2(\mathcal{G}, \mathbb{C}^2) : \int_{\sigma(\mathcal{D}_\mathcal{G})} |\nu| \, d\mu^{\psi}_{\mathcal{D}_\mathcal{G}}(\nu) \right\}, \qquad \mathcal{Q}_{\mathcal{D}_\mathcal{G}}(\psi) := \int_{\sigma(\mathcal{D}_\mathcal{G})} \nu \, d\mu^{\psi}_{\mathcal{D}_\mathcal{G}}(\nu),$$

where $\mu^{\psi}_{\mathcal{D}_\mathcal{G}}$ denotes the spectral measure associated with $\mathcal{D}_\mathcal{G}$ and ψ. However, such a definition is not useful for computations. A more precise description of the quadratic form and its domain can be obtained arguing as follows (for more details, see [39]).

First, using Real Interpolation Theory, one can prove that

$$\mathrm{dom}(\mathcal{Q}_{\mathcal{D}_\mathcal{G}}) = \left[L^2(\mathcal{G}, \mathbb{C}^2), \mathrm{dom}(\mathcal{D}_\mathcal{G}) \right]_{\frac{1}{2}}, \tag{25}$$

whence

$$\mathrm{dom}(\mathcal{Q}_{\mathcal{D}_\mathcal{G}}) \hookrightarrow H^{1/2}(\mathcal{G}, \mathbb{C}^2) \hookrightarrow L^p(\mathcal{G}, \mathbb{C}^2), \qquad \forall p \geq 2.$$

On the other hand, according to (24) one can decompose the form domain as the orthogonal sum of the positive and negative *spectral subspaces* for the operator $\mathcal{D}_\mathcal{G}$, i.e.,

$$\mathrm{dom}(\mathcal{Q}_{\mathcal{D}_\mathcal{G}}) = \mathrm{dom}(\mathcal{Q}_{\mathcal{D}_\mathcal{G}})^+ \oplus \mathrm{dom}(\mathcal{Q}_{\mathcal{D}_\mathcal{G}})^-.$$

As a consequence, if one denotes $\psi^+ := P^+\psi$ and $\psi^- := P^-\psi$ and recalls that it is possible to define a norm for $\mathrm{dom}(\mathcal{Q}_{\mathcal{D}_\mathcal{G}})$ as

$$\|\psi\|_{\mathcal{Q}_{\mathcal{D}_\mathcal{G}}} := \|\sqrt{|\mathcal{D}_\mathcal{G}|}\psi\|_{2,\mathcal{G}}, \qquad \forall \psi \in \mathrm{dom}(\mathcal{Q}_{\mathcal{D}_\mathcal{G}}),$$

(with $\sqrt{|\mathcal{D}_\mathcal{G}|}\psi$ given by Borel functional calculus for $\mathcal{D}_\mathcal{G}$), then there results that

$$\mathcal{Q}_{\mathcal{D}_\mathcal{G}}(\psi) = \frac{1}{2}\left(\|\psi^+\|_{\mathcal{Q}_{\mathcal{D}_\mathcal{G}}} - \|\psi^-\|_{\mathcal{Q}_{\mathcal{D}_\mathcal{G}}}\right). \tag{26}$$

Clearly, (25) and (26) are not explicit forms for the domain and the action of the quadratic form associated with $\mathcal{D}_\mathcal{G}$. Nevertheless, they present the suitable details to manage the computations required in the proofs of Theorems 6 and 7 below.

3.2. Bound States

The study of bound states of (7) presents some relevant differences with respect to the NLS case (5) (a difference which also arises in the extended case).

The main point here is the unboundedness from below of the spectrum of $\mathcal{D}_\mathcal{G}$ which makes the associated quadratic form *strongly indefinite*, even fixing the L^2-norm. Consequently, the natural energy functional associated with (7), i.e.,

$$\mathcal{Q}_{\mathcal{D}_\mathcal{G}}(\psi) - \frac{1}{p}\int_\mathcal{K} |\psi|^{p-2}\psi, \qquad p \geq 2,$$

is unbounded from below even under the mass constraint $\|\psi\|^2_{2,\mathcal{G}} = \mu$. Hence, no minimization can be performed and the search for constrained critical points presents several technical difficulties.

Therefore, the most promising strategy seems to be that of considering the *action functional* associated with (7) without any constraint. Then, for a fixed $\omega \in \mathbb{R}$, one looks for critical points of the functional

$$\mathcal{L}(\psi, \mathcal{K}, p) = \mathcal{Q}_{\mathcal{D}_\mathcal{G}}(\psi) - \frac{\omega}{2}\int_\mathcal{G} |\psi|^2 - \frac{1}{p}\int_\mathcal{K} |\psi|^p\,dx,$$

on the form domain $\mathrm{dom}(\mathcal{Q}_{\mathcal{D}_\mathcal{G}})$.

To this aim, one needs to adapt to the metric graphs setting some well-known techniques of Critical Point Theory. In particular, a suitable version of the so-called *linking* technique for even functional is exploited. We refer the reader to [51,55], for a general presentation of those methods in an abstract setting, together with several applications, and to [56] which deals with NLDEs.

Critical levels for $\mathcal{L}$ are defined as the following min-max levels

$$\alpha_N := \inf_{X \in \mathcal{F}_N} \sup_{\psi \in X} \mathcal{L}(\psi, \mathcal{K}, p),$$

with

$$\mathcal{F}_N := \left\{X \in H^1(\mathcal{G}, \mathbb{C}^2)\backslash\{0\} : X \text{ closed and symmetric, s.t. } \gamma[h_t(S_r^+) \cap X] \geq N,\ \forall t \geq 0\right\},$$

where h_t is the usual *pseudo-gradient flow* associated with $\mathcal{L}(\cdot, \mathcal{K}, p)$, S_r^+ is the sphere of radius r of $\mathrm{dom}(\mathcal{Q}_{\mathcal{D}_\mathcal{G}})^+$ and γ denotes again the Krasnosel'skii genus. Indeed, one can check that if $\mathcal{F}_N \neq \emptyset$, then there exists a Palais-Smale sequence $(\psi_n) \subset \mathrm{dom}(\mathcal{Q}_{\mathcal{D}_\mathcal{G}})$ at level α_N, i.e.,

$$\begin{cases} \mathcal{L}(\psi_n, \mathcal{K}, p) \to \alpha_N \\ d\mathcal{L}(\psi_n, \mathcal{K}, p) \xrightarrow{\mathrm{dom}(\mathcal{Q}_{\mathcal{D}_\mathcal{G}})'} 0, \end{cases}$$

and in addition, there results

$$\alpha_{N_1} \leq \alpha_{N_2}, \qquad \forall N_1 < N_2,$$

$$\rho \leq \alpha_N \leq \widetilde{\rho} < +\infty, \qquad \forall N \in \mathbb{N},$$

for fixed $\rho, \tilde{\rho} > 0$.

As a consequence, in order to prove that $\mathcal{F}_N \neq \emptyset$, it suffices to show that the functional possesses a so-called *linking geometry*; namely that for every $N \in \mathbb{N}$ there exist $R = R(N,p) > 0$ and an N-dimensional space $Z_N \subset \mathrm{dom}(\mathcal{Q}_{\mathcal{D}_\mathcal{G}})^+$ such that

$$\mathcal{L}(\psi, \mathcal{K}, p) \leq 0, \qquad \forall \psi \in \partial \mathcal{M}_N,$$

where

$$\partial \mathcal{M}_N = \{\psi \in \mathcal{M}_N : \|\psi^-\| = R \text{ or } \|\psi^+\| = R\}$$

and

$$\mathcal{M}_N := \left\{\psi \in \mathrm{dom}(\mathcal{Q}_{\mathcal{D}_\mathcal{G}}) : \|\psi^-\| \leq R \text{ and } \psi^+ \in Z_N \text{ with } \|\psi^+\| \leq R\right\},$$

and that there exist $r, \rho > 0$ such that

$$\inf_{S_r^+} \mathcal{L}(\cdot, \mathcal{K}, p) \geq \rho > 0$$

(for a graphic intuition see, e.g., Figure 10).

Finally, checking the validity of the Palais-Smale condition at positive levels for the action functional it is possible to prove the following

Theorem 6 ([39] [Theorem 2.11]). *Let $\mathcal{G}$ satisfy* **(H1)**–**(H3)**, *$p > 2$ and $m, c > 0$. Then, for every $\omega \in \mathbb{R} \backslash \sigma(\mathcal{D}_\mathcal{G}) = (-mc^2, mc^2)$ there exists infinitely many (distinct) pairs of bound states of frequency ω of the NLDE, at mass m and relativistic parameter c.*

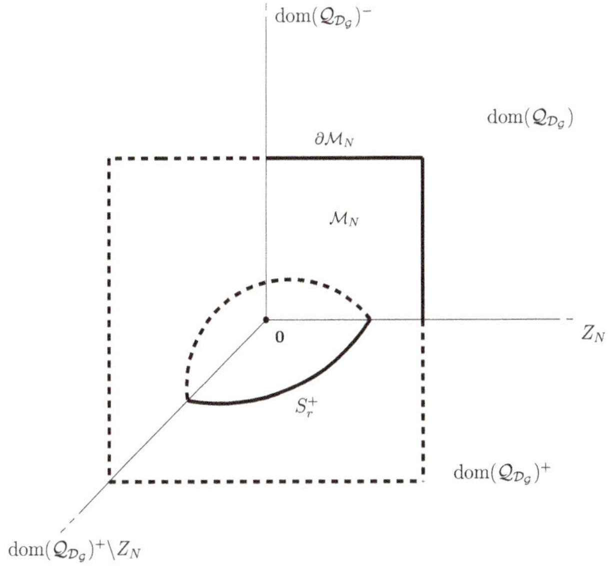

Figure 10. A graphic insight on linking geometry.

4. Nonrelativistic Limit

The NLDE and the NLSE equation are clearly closely related as, physically, the latter should correspond to the nonrelativistic limit of the former. Heuristically, one expects to recover the NLSE from NLDE as the relativistic parameter tends to infinity (namely, "$c \to \infty$"), that is, neglecting relativistic effects. It is, then, particularly interesting to rigorously establish such a connection.

This has been first done in [57] for the Dirac-Fock equations proving the convergence of bound states to those of the (nonrelativistic) Hartree-Fock model.

More in detail, given a sequence $c_n \to \infty$, one is interested in the limit behavior of a sequence of bound states with frequencies $0 < \omega_n < mc_n^2$. Precisely, one must choose frequencies such that

$$\omega_n - mc_n^2 \longrightarrow \frac{\lambda}{m} < 0.$$

This represents a kind of renormalization, which corresponds to the fact that one must subtract the *rest energy* of the particles, as this property is a peculiarity of relativistic theories which is absent in the nonrelativistic setting.

The strategy adopted in [39] is an adaptation of the one developed in [57]. Namely, first one has to establish H^1 uniform bounds for the sequence of the bound states and then, suitably manipulating (7), one has to prove that the lower component goes to zero while the sequence of upper components is a Palais-Smale sequence for the action functional associated with (5), namely for

$$\mathcal{J}(u) := \frac{1}{2}\int_{\mathcal{G}} |u'|^2\,dx - \frac{2m}{p}\int_{\mathcal{K}} |u|^p\,dx - \frac{\lambda}{2}\int_{\mathcal{G}} |u|^2\,dx.$$

at a fixed $\lambda < 0$. Therefore, up to a compactness argument, the existence of a limit satisfying the NLSE is proved.

However, we would like to remark that some relevant differences are present here with respect to [57] that call for some relevant modifications of the proofs. In particular, in [57] bound states were obtained as constrained critical points of the energy. Consequently, the uniform boundedness of the L^2-norms and the non-triviality of the limit are easily obtained. On the contrary, in the case considered in [39], those properties are not a priori guaranteed and represent two main points of the proofs.

However, for the sake of completeness, we mention that in [57] one of the most delicate parts the proof is to estimate the sequence of the Lagrange multipliers corresponding to the L^2-constraint, proving that they are in the spectral gap of the operator. This, actually, is one of the reasons for which in [39] we looked for critical points of the action and not for the constrained critical points of the energy, to avoid such additional technical difficulties.

In view of all the above remarks it is possible to state the following

Theorem 7 ([39] [Theorem 2.12]). *Let $\mathcal{G}$ satisfy* **(H1)**–**(H3)**, *$p \in (2,6)$, $m > 0$ and $\lambda < 0$. Let also (c_n), (ω_n) be two real sequences such that*

$$0 < c_n, \omega_n \to +\infty,$$

$$\omega_n < mc_n^2,$$

$$\omega_n - mc_n^2 \longrightarrow \frac{\lambda}{m}.$$

If $\{\psi_n = (\phi_n, \chi_n)^T\}$ is a sequence of bound states of frequency ω_n of the NLDE at relativistic parameter c_n, then, up to subsequences, there holds

$$\phi_n \to u \qquad \text{and} \qquad \chi_n \to 0 \qquad \text{in} \quad H^1(\mathcal{G}).$$

where u is a bound state of frequency λ of the NLSE.

Actually, the bound states of NLDE do not converge exactly to the bound states of the NLSE depicted in (7). In fact, u is a solution of

$$-\Delta_{\mathcal{G}} u - 2m\chi_{\mathcal{K}}|u|^{p-2}\, u = \lambda u.$$

The coefficient $2m$ is consistent with the fact that in the nonrelativistic limit the kinetic (free) part of the Hamiltonian of a particle is given by

$$\frac{-\Delta}{2m}.$$

Remark 3. *It is also worth observing that Theorem 7, in contrast to Theorem 6, holds only for a fixed range of exponents, the L^2-subcritical case: $p \in (2,6)$.*

5. Conclusion: A Brief Summary

In this final section we provide in Table 1 a summary of the results presented in the review. However, before showing it, some remarks are in order:

(i) the mentioned results only concerne the "free" self-adjoint extensions of the Laplacian and the Dirac operator introduced in Sections 1 and 3.1, respectively;

Table 1. Summarizing Table.

	Exponents	Ground	Bound	Connection NLS-NLD
NLSE	$p \in (2,4)$	- yes, $\forall \mu > 0$	(see box below)	(see box below)
	$p \in [4,6)$	- yes if $\mu > \mu_1$ - no if $\mu < \mu_2$ - unknown if $\mu \in [\mu_2, \mu_1]$	- yes (and multiple) if μ is large enough - yes if $\mathcal{G}$ has a loop or two terminal edges - no (with $\lambda \leq 0$) if $\mu < (p/2)^{2/(2-p)}\mu_2$ - no (with $\lambda \geq 0$) if $\mathcal{G}$ has (at most) one terminal edge and no loops - unknown otherwise	- some bounds are limits of bounds of NLDE - unknown if all are limits
	$p = 6$	- yes if $\mu \in [\mu_{\mathcal{K}}, \mu_{\mathbb{R}}]$ and if no terminal edges and no cycle-coverings* - no otherwise	- yes if $\mathcal{G}$ has a loop or two terminal edges - no (with $\lambda \geq 0$) if $\mathcal{G}$ has (at most) one terminal edge and no loops - unknown otherwise	- unknown
	$p > 6$	- unknown	(see box above)	- unknown
NLDE	$p \in (2,6)$	- no	- yes (infinitely many) if $\omega \in (-mc^2, mc^2)$, namely $\omega \in \mathbb{R}\backslash\sigma(\mathcal{D}_{\mathcal{G}})$ - unknown otherwise	- bounds converge (up to subsequences) to bounds of NLSE - unknown if can one avoid extracting
	$p = 6$	- no	(see box above)	- unknown
	$p > 6$	- no	(see box above)	- unknown

(ii) in order to find bound states, in the NLS case one fixes the mass μ and studies constrained critical points of the energy functional (thus providing no information on the frequencies λ that arise naturally as Lagrange multipliers), while in the NLD case one fixes the frequency ω and discusses the connected action functional (thus losing any information on the mass of the resulting critical points);

(iii) the nonrelativistic limit must be considered (as above) a limit for a sequence of relativistic parameters $c_n \to \infty$ and a suitably "tuned" sequence of frequencies ω_n;
(iv) since (clearly) a ground state is a bound state too, the fourth column of the Table 1 must be meant to refer to those bound states which are not ground states;
(v) the constants μ_1, μ_2 are defined by: $\mu_1 := N^{\frac{4}{p-2}} c_p^{\frac{6-p}{p-2}} |\mathcal{K}|^{\frac{p-6}{p-2}}$ (with N, c_p introduced in Theorem 1–item (i)) and $\mu_2 := (p/2)^{\frac{2}{p-2}} C(\mathcal{G}, p)^{\frac{4-p}{p-2}} |\mathcal{K}|^{\frac{p-6}{p-2}} C(\mathcal{G}, \infty)^{\frac{p(p-6)}{p-2}}$ (with $C(\mathcal{G}, p)$, $C(\mathcal{G}, \infty)$ introduced in (10)–(11));
(v) the constant $\mu_{\mathcal{K}}$ (defined by (19)), in contrast to $\mu_{\mathbb{R}}$ (defined by (18)), actually depends on the graph and, moreover, if there is more than one half-line, then existence is guaranteed only provided that $\mu_{\mathcal{K}} \neq \mu_{\mathbb{R}}$.

Author Contributions: All authors contributed equally to this work.

Funding: This research was partially funded by *INdAM—GNAMPA Project 2018: "Variational Problems and Applications"*.

Conflicts of Interest: The authors declare no conflict of interest.

References

1. Adami, R.; Serra, E.; Tilli, P. NLS ground states on graphs. *Calc. Var. Part. Differ. Equ.* **2015**, *54*, 743–761. [CrossRef]
2. Berkolaiko, G.; Kuchment, P. *Introduction to Quantum Graphs*; Mathematical Surveys and Monographs, 186; American Mathematical Society: Providence, RI, USA, 2013; ISBN 978-0-8218-9211-4.
3. Adami, R.; Serra, E.; Tilli, P. Nonlinear dynamics on branched structures and networks. *Riv. Math. Univ. Parma (N.S.)* **2017**, *8*, 109–159.
4. Gnutzmann, S.; Waltner, D. Stationary waves on nonlinear quantum graphs: General framework and canonical perturbation theory. *Phys. Rev. E* **2016**, *93*, 032204. [CrossRef] [PubMed]
5. Noja, D. Nonlinear Schrödinger equation on graphs: Recent results and open problems. *Philos. Trans. R. Soc. Lond. Ser. A Math. Phys. Eng. Sci.* **2014**, *372*, 20130002. [CrossRef]
6. Lorenzo, M.; Lucci, M.; Merlo, V.; Ottaviani, I.; Salvato, M.; Cirillo, M.; Müller, F.; Weimann, T.; Castellano, M.G.; Chiarello, F.; et al. On Bose-Einstein condensation in Josephson junctions star graph arrays. *Phys. Lett. A* **2014**, *378*, 655–658. [CrossRef]
7. Adami, R.; Cacciapuoti, C.; Finco, D.; Noja, D. Fast solitons on star graphs. *Rev. Math. Phys.* **2011**, *23*, 409–451. [CrossRef]
8. Adami, R.; Cacciapuoti, C.; Finco, D.; Noja, D. On the structure of critical energy levels for the cubic focusing NLS on star graphs. *J. Phys. A* **2012**, *45*, 3738–3777. [CrossRef]
9. Adami, R.; Cacciapuoti, C.; Finco, D.; Noja, D. Variational properties and orbital stability of standing waves for NLS equation on a star graph. *J. Differ. Equ.* **2014**, *257*, 3738–3777. [CrossRef]
10. Adami, R.; Cacciapuoti, C.; Finco, D.; Noja, D. Constrained energy minimization and orbital stability for the NLS equation on a star graph. *Ann. Inst. H Poincaré Anal. Non Linéaire* **2014**, *31*, 1289–1310. [CrossRef]
11. Adami, R.; Serra, E.; Tilli, P. Threshold phenomena and existence results for NLS ground states on metric graphs. *J. Funct. Anal.* **2016**, *271*, 201–223. [CrossRef]
12. Adami, R.; Serra, E.; Tilli, P. Negative energy ground states for the L2-critical NLSE on metric graphs. *Commun. Math. Phys.* **2017**, *352*, 387–406. [CrossRef]
13. Adami, R.; Serra, E.; Tilli, P. Multiple positive bound states for the subcritical NLS equation on metric graphs. *Calc. Var. Part. Diff. Equ.* **2019**, *58*. [CrossRef]
14. Cacciapuoti, C.; Finco, D.; Noja, D. Topology-induced bifurcations for the nonlinear Schrödinger equation on the tadpole graph. *Phys. Rev. E* **2015**, *91*, 013206. [CrossRef] [PubMed]
15. Kairzhan, A.; Pelinovsky, D.E. Nonlinear instability of half-solitons on star graphs. *J. Differ. Equ.* **2018**, *264*, 7357–7383. [CrossRef]
16. Noja, D.; Pelinovsky, D.; Shaikhova, G. Bifurcations and stability of standing waves in the nonlinear Schrödinger equation on the tadpole graph. *Nonlinearity* **2015**, *28*, 2343–2378. [CrossRef]

17. Noja, D.; Rolando, S.; Secchi, S. Standing waves for the NLS on the double-bridge graph and a rational-irrational dichotomy. *J. Differ. Equ.* **2019**, *266*, 147–178. [CrossRef]
18. Cacciapuoti, C.; Finco, D.; Noja, D. Ground state and orbital stability for the NLS equation on a general starlike graph with potentials. *Nonlinearity* **2017**, *30*, 3271–3303. [CrossRef]
19. Gnutzmann, S.; Smilansky, U.; Derevyanko, S. Stationary scattering from a nonlinear network. *Phys. Rev. A* **2011**, *83*, 033831. [CrossRef]
20. Serra, E.; Tentarelli, L. Bound states of the NLS equation on metric graphs with localized nonlinearities. *J. Differ. Equ.* **2016**, *260*, 5627–5644. [CrossRef]
21. Serra, E.; Tentarelli, L. On the lack of bound states for certain NLS equations on metric graphs. *Nonlinear Anal.* **2016**, *145*, 68–82. [CrossRef]
22. Tentarelli, L. NLS ground states on metric graphs with localized nonlinearities. *J. Math. Anal. Appl.* **2016**, *433*, 291–304. [CrossRef]
23. Dovetta, S.; Tentarelli, L. Ground states of the L^2-critical NLS equation with localized nonlinearity on a tadpole graph. *arXiv* **2018**, arXiv:1803.09246.
24. Dovetta, S.; Tentarelli, L. L^2-critical NLS on noncompact metric graphs with localized nonlinearity: Topological and metric features. *arXiv* **2018**, arXiv:1811.02387.
25. Cacciapuoti, C.; Dovetta, S.; Serra, E. Variational and stability properties of constant solutions to the NLS equation on compact metric graphs. *Milan J. Math.* **2018**, *86*, 305–327. [CrossRef]
26. Dovetta, S. Existence of infinitely many stationary solutions of the L^2-subcritical and critical NLSE on compact metric graphs. *J. Differ. Equ.* **2018**, *264*, 4806–4821. [CrossRef]
27. Duca, A. Global exact controllability of the bilinear Schrödinger potential type models on quantum graphs. *arXiv* **2017**, arXiv:1710.06022.
28. Marzuola, J.L.; Pelinovsky, D.E. Ground state on the dumbbell graph. *Appl. Math. Res. Express AMRX* **2016**, 98–145. [CrossRef]
29. Adami, R.; Dovetta, S.; Serra, E.; Tilli, P. Dimensional crossover with a continuum of critical exponents for NLS on doubly periodic metric graphs. *arXiv* **2018**, arXiv:1805.02521.
30. Dovetta, S. Mass-constrained ground states of the stationary NLSE on periodic metric graphs. *arXiv* **2018**, arXiv:1811.06798.
31. Gilg, S.; Pelinovsky, D.E.; Schneider, G. Validity of the NLS approximation for periodic quantum graphs. *NoDEA Nonlinear Differ. Equ. Appl.* **2016**, *23*, 63. [CrossRef]
32. Pelinovsky, D.E.; Schneider, G. Bifurcations of standing localized waves on periodic graphs. *Ann. Henri Poincaré* **2017**, *18*, 1185–1211. [CrossRef]
33. Mugnolo, D.; Noja, D.; Seifert, C. Airy-type evolution equations on star graphs. *Anal. PDE* **2018**, *11*, 1625–1652. [CrossRef]
34. Musslimani, Z.H.; Makris, K.G.; El-Ganainy, R.; Christodoulides, D.N. Optical solitons in PT periodic potentials. *Phys. Rev. Lett.* **2008**, *100*, 030402. [CrossRef] [PubMed]
35. Ablowitz, M.J.; Musslimani, Z.H. Integrable nonlocal nonlinear Schrödinger equation. *Phys. Rev. Lett.* **2013**, *110*, 064105. [CrossRef] [PubMed]
36. Ablowitz, M.J.; Musslimani, Z.H. Integrable nonlocal nonlinear equation. *Stud. Appl. Math.* **2017**, *139*, 7–59. [CrossRef]
37. Kurasov, P.; Majidzadeh Garjani, B. Quantum graphs: PT -symmetry and reflection symmetry of the spectrum. *J. Math. Phys.* **2017**, *58*, 023506. [CrossRef]
38. Matrasulov, D.U.; Sabirov, K.K.; Yusupov J.R.; PT-symmetric quantum graphs. *arXiv* **2018**, arXiv:1805.08104.
39. Borrelli, W.; Carlone, R.; Tentarelli, L. Nonlinear Dirac equation on graphs with localized nonlinearities: Bound states and nonrelativistic limit. *arXiv* **2018**, arXiv:1807.06937.
40. Haddad, L.H.; Carr, L.D. The nonlinear Dirac equation in Bose-Einstein condensates: Foundation and symmetries. *Physical D* **2009**, *238*, 1413–1421. [CrossRef]
41. Tran, T.X.; Longhi, S.; Biancalana, F. Optical analogue of relativistic Dirac solitons in binary waveguide arrays. *Ann. Phys.* **2014**, *340*, 179–187. [CrossRef]
42. Arbunich, J.; Sparber, C. Rigorous derivation of nonlinear Dirac equations for wave propagation in honeycomb structures. *J. Math. Phys.* **2018**, *59*, 011509. [CrossRef]
43. Borrelli, W. Stationary solutions for the 2D critical Dirac equation with Kerr nonlinearity. *J. Differ. Equ.* **2017**, *263*, 7941–7964. [CrossRef]

44. Borrelli, W. Multiple solutions for a self-consistent Dirac equation in two dimensions. *J. Math. Phys.* **2018**, *59*, 041503. [CrossRef]
45. Borrelli, W. Weakly localized states for nonlinear Dirac equations. *Calc. Var. Part. Differ. Equ.* **2018**, *57*, 155. [CrossRef]
46. Fefferman, C.L.; Weinstein, M.I. Wave Packets in Honeycomb Structures and Two-Dimensional Dirac Equations. *Commun. Math. Phys.* **2014**, *326*, 251–286. [CrossRef]
47. Sabirov, K.K.; Babajanov, D.B.; Matrasulov, D.U.; Kevrekidis, P.G. Dynamics of Dirac solitons in networks. *J. Phys. A* **2018**, *51*, 435203. [CrossRef]
48. Cazenave, T. *Semilinear Schrödinger Equations*; Courant Lecture Notes in Mathematics, 10; American Mathematical Society: Providence, RI, USA, 2003, ISBN 0-8218-3399-5.
49. Ambrosetti, A.; Malchiodi, A. *Nonlinear Analysis and Semilinear Elliptic Problems*; Cambridge Studies in Advanced Mathematics, 104; Cambridge University Press: Cambridge, UK, 2007, ISBN 978-0-521-86320-9.
50. Berestycki, H.; Lions, P.-L. Nonlinear scalar field equations II. Existence of infinitely many solutions. *Arch. Rational. Mech. Anal.* **1983**, *82*, 347–375. [CrossRef]
51. Rabinowitz, P.H. *Minimax Methods in Critical Point Theory With Applications to Differential Equations*; CBMS Regional Conference Series in Mathematics, 65; American Mathematical Society: Providence, RI, USA, 1986.
52. Krasnosel'skii, M. A. *Topological Methods in the Theory of Nonlinear Integral Equations*; A Pergamon Press Book; The Macmillan Co.: New York, NY, USA, 1964.
53. Bulla, W.; Trenkler, T. The free Dirac operator on compact and noncompact graphs. *J. Math. Phys.* **1990**, *31*, 1157–1163. [CrossRef]
54. Post, O. Equilateral quantum graphs and boundary triples. In *Analysis on Graphs and Its Applications*; Proc. Sympos. Pure Math., 77; American Mathematical Society: Providence, RI, USA, 2008; pp. 469–490.
55. Struwe, M. *Variational Methods. Applications to Nonlinear Partial Differential Equations and Hamiltonian Systems*, 4th ed.; Results in Mathematics and Related Areas, 3rd Series; A Series of Modern Surveys in Mathematics, 34; Springer-Verlag: Berlin, Germany, 2008, ISBN 978-3-540-74012-4.
56. Esteban, M.J.; Séré, E. Stationary states of the nonlinear Dirac equation: A variational approach. *Commun. Math. Phys.* **1995**, *171*, 323–350. [CrossRef]
57. Esteban, M.J.; Séré, E. Nonrelativistic limit of the Dirac-Fock equations. *Ann. Henri Poincaré* **2001**, *2*, 941–961. [CrossRef]

symmetry

MDPI

Article

A Note on Sign-Changing Solutions to the NLS on the Double-Bridge Graph

Diego Noja *, Sergio Rolando and Simone Secchi

Dipartimento di Matematica e Applicazioni, Università di Milano Bicocca, via R. Cozzi 55, 20126 Milano, Italy; sergio.rolando@unimib.it (S.R.); simone.secchi@unimib.it (S.S.)

* Correspondence: diego.noja@unimib.it

Received: 15 January 2019; Accepted: 28 January 2019; Published: 1 February 2019

Abstract: We study standing waves of the NLS equation posed on the double-bridge graph: two semi-infinite half-lines attached at a circle. At the two vertices, Kirchhoff boundary conditions are imposed. We pursue a recent study concerning solutions nonzero on the half-lines and periodic on the circle, by proving some existing results of sign-changing solutions non-periodic on the circle.

Keywords: quantum graphs; non-linear Schrödinger equation; standing waves

MSC: 35Q55; 81Q35; 35R02

1. Introduction and Main Results

The study of nonlinear equations on graphs, especially the nonlinear Schrödinger equation (NLS), is a quite recent research subject, which already produced a plenty of interesting results (see [1–3]). The attractive feature of these mathematical models is the complexity allowed by the graph structure, joined with the one dimensional character of the equations. While they are an oversimplification in many real problems, they appear indicative of several dynamically interesting phenomena that are atypical or unexpected in more standard frameworks. The most studied issue concerning NLS is certainly the existence and characterization of standing waves (see, e.g., [4–9]). More particularly, several results are known about ground states (standing waves of minimal energy at fixed mass, i.e., L^2 norm) as regard existence, non-existence and stability properties, depending on various characteristics of the graph [2,10–13].

In this paper, we are interested in a special example, which reveals an unsuspectedly complex structure of the set of standing waves. More precisely, we consider a metric graph $\mathcal{G}$ made up of two half lines joined by two bounded edges, i.e., a so-called double-bridge graph (see Figure 1). $\mathcal{G}$ can also be thought of as a ring with two half lines attached in two distinct vertices. The half lines are both identified with the interval $[0, +\infty)$, while the bounded edges are represented by two bounded intervals of lengths $L_1 > 0$ and $L_2 \geq L_1$, precisely $[0, L_1]$ and $[L_1, L]$ with $L = L_1 + L_2$.

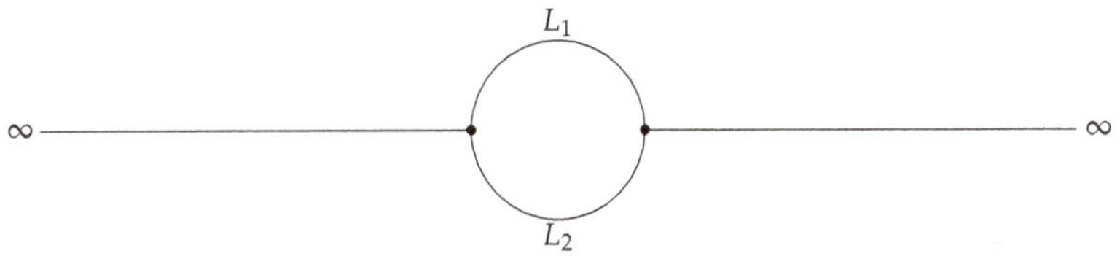

Figure 1. The double-bridge graph.

A function ψ on $\mathcal{G}$ is a Cartesian product $\psi(x_1, ..., x_4) = (\psi_1(x_1), \ldots, \psi_4(x_4))$ with $x_j \in I_j$ for $j = 1, \ldots, 4$, where $I_1 = [0, L_1]$, $I_2 = [L_1, L]$ and $I_3 = I_4 = [0, +\infty)$. Then, a Schrödinger operator $H_{\mathcal{G}}$ on $\mathcal{G}$ is defined as

$$H_{\mathcal{G}}\psi(x_1, \ldots, x_4) = (-\psi_1''(x_1), \ldots, -\psi_4''(x_4)), \quad x_j \in I_j, \tag{1}$$

with domain $D(H_{\mathcal{G}})$ given by the functions ψ on $\mathcal{G}$ whose components satisfy $\psi_j \in H^2(I_j)$ together with the so-called *Kirchhoff boundary conditions*, i.e.,

$$\psi_1(0) = \psi_2(L) = \psi_3(0), \quad \psi_1(L_1) = \psi_2(L_1) = \psi_4(0), \tag{2}$$

$$\psi_1'(0) - \psi_2'(L) + \psi_3'(0) = \psi_1'(L_1) - \psi_2'(L_1) - \psi_4'(0) = 0. \tag{3}$$

As is well known (see [14] for general information on quantum graphs), the operator $H_{\mathcal{G}}$ is self-adjoint on the domain $D(H_{\mathcal{G}})$, and it generates a unitary Schrödinger dynamics. Essential information about its spectrum is given in ([15], Appendix A). We perturb this linear dynamics with a focusing cubic term, namely we consider the following NLS on $\mathcal{G}$

$$i\frac{d\psi_t}{dt} = H_{\mathcal{G}}\psi_t - |\psi_t|^2\psi_t \tag{4}$$

where the nonlinear term $|\psi_t|^2\psi_t$ is a shortened notation for $(|\psi_{1,t}|^2\psi_{1,t}, \ldots, |\psi_{4,t}|^2\psi_{4,t})$. Hence, Equation (4) is a system of scalar NLS equations on the intervals I_j coupled through the Kirchhoff boundary conditions in Equations (2)–(3) included in the domain of $H_{\mathcal{G}}$. On rather general grounds, it can be shown that this problem enjoys well-posedness both in strong sense and in the energy space (see in particular ([2], Section 2.6)).

We are interested in standing waves of Equation (4), i.e., its solutions of the form $\psi_t = e^{-i\omega t}U(x)$ where $\omega \in \mathbb{R}$ and $U(x_1, ..., x_4) = (u_1(x_1), \ldots, u_4(x_4))$ is a purely spatial function on $\mathcal{G}$, which may also depend on ω. Such a problem has already been considered in [11,12,15,16]. In particular, in [11,12], variational methods are used to show, among many other things, that Equation (4) has no ground state, i.e., no standing wave exists that minimizes the energy at fixed L^2-norm. In a recent paper [16], information on *positive* bound states that are not ground states is given. The special example of tadpole graph (a ring with a single half-line) is treated in detail in [17,18].

As for the results in [15], they can be summarized as follows. Writing the problem of standing waves of Equation (4) component-wise, we get the following scalar problem:

$$\begin{cases} -u_j'' - u_j^3 = \omega u_j, & u_j \in H^2(I_j) \\ u_1(0) = u_2(L) = u_3(0), & u_1(L_1) = u_2(L_1) = u_4(0) \\ u_1'(0) - u_2'(L) + u_3'(0) = 0, & u_1'(L_1) - u_2'(L_1) - u_4'(0) = 0. \end{cases} \tag{5}$$

Such a system has solutions with $u_3 = u_4 = 0$ if and only if the ratio L_1/L_2 is rational. In this case, they form a sequence of continuous branches in the $(\omega, \|U\|_{L^2})$ plane, bifurcating from the linear eigenvectors of the Schrödinger operator $H_{\mathcal{G}}$ (see Figure 2), and they are periodic on the ring of $\mathcal{G}$, that is, u_1 and u_2 are restrictions to I_1 and I_2 of a function u belonging to the second Sobolev space of periodic functions $H^2_{\text{per}}([0, L]) = \{u \in H^2([0, L]) : u(0) = u(L),\ u'(0) = u'(L)\}$. In particular, such function u is a rescaled Jacobi cnoidal function (see, e.g., [19,20] for a treatise on the Jacobian elliptic functions). If $\omega \geq 0$, no other nonzero standing waves exist, since the NLS on the unbounded edges has no nontrivial solution. If $\omega < 0$, instead, the NLS on the half lines has soliton solutions, so that standing waves with nonzero u_3 and u_4 are admissible. The general study of this kind of solutions leads to a rather complicated system of equations, since, while u_3 and u_4 must be shifted solitons, each of u_1 and u_2 can be (at least in principle) a cnoidal function, a dnoidal function or a shifted soliton. To limit this complexity, the analysis in [15] is focused on the special case of standing waves that are non-vanishing on the half lines but share the above-mentioned periodicity feature with the bifurcation solutions. This amounts to study the following system:

$$\begin{cases} -u'' - u^3 = \omega u, & u \in H^2_{\text{per}}([0,L]),\ \omega < 0 \\ u(0) = \pm u(L_1) = \sqrt{2|\omega|} \end{cases} \tag{6}$$

where the sign $\pm$ distinguishes the cases of u_3 and u_4 with the same sign (which we may assume positive, thanks to the odd parity of the equation) or with different signs. In [15], it is shown that:

(i) If $L_1/L_2 \in \mathbb{Q}$, then the set of solutions to (6) is made up of a sequence of secondary bifurcation branches $\{(\omega, \tilde{u}_{n,\omega}) : \omega < 0\}_{n\geq 1}$, originating at $\omega = 0$ from each of the previous ones, together with a sequence $\{(\omega_n, u_n)\}_{n\geq 1}$ not lying on any branch (see Figure 2).

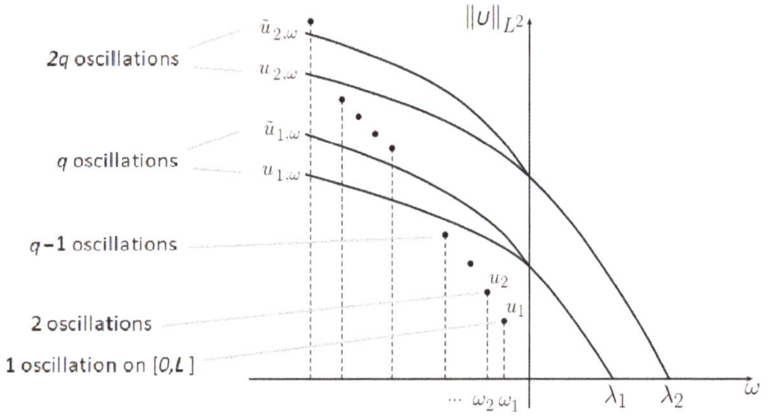

Figure 2. Bifurcation diagram for $L_1/L = p/q$ with $p, q \in \mathbb{N}$ coprime.

(ii) If $L_1/L_2 \notin \mathbb{Q}$, then the set of solutions to (6) reduces to two sequences $\{(\omega_n^+, u_n^+)\}_{n\geq 1}$ and $\{(\omega_n^-, u_n^-)\}_{n\geq 1}$ alone, solving the problem in Equation (6) with sign $\pm$, respectively, where the frequency sequences $\{\omega_n^\pm\}_{n\geq 1}$ are unbounded below and have at least a finite nonzero cluster point (see Figure 3). The functions $u_n^\pm$ oscillate n times on the ring of the graph.

These results come rather unexpectedly, so the aim of this paper is to pursue the study begun in [15] by deepening the understanding of such results in relation to the underlying physical model. In particular, we ask the following questions: Does Equation (4) admit standing waves that are non-periodic on ring of $\mathcal{G}$? If so, do they form continuous branches to which the isolated periodic solutions belong?

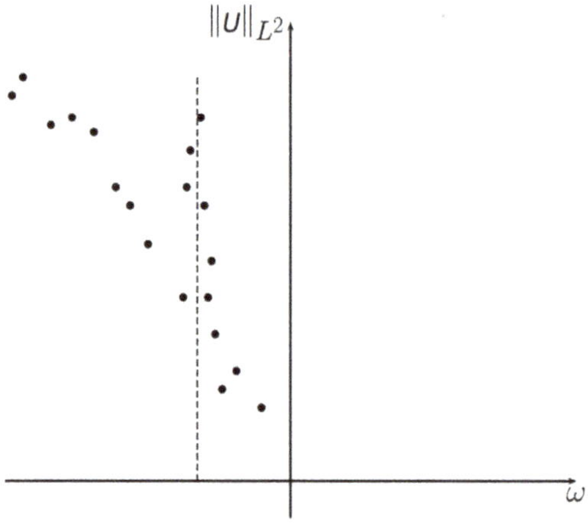

Figure 3. The appearance of each of the sequences $\{(\omega_n^+, u_n^+)\}_{n\in\mathbb{N}}$ and $\{(\omega_n^-, u_n^-)\}_{n\in\mathbb{N}}$ for $L_1/L \in \mathbb{R}\setminus\mathbb{Q}$.

With a view to especially answer the second question, we look for standing waves which include the ones given by Equation (6) but still change sign on the bounded edges. More precisely, we look for solutions to Equation (5) exhibiting the following features:

- u_1, u_2 are sign-changing.
- u_3, u_4 are nonzero.

The second feature implies $\omega < 0$ and

$$u_j(x) = \pm\sqrt{2}\eta \operatorname{sech}\left(\eta\left(x + a_j\right)\right), \qquad a_j \in \mathbb{R},\ j = 3,4 \tag{7}$$

where we set $\eta := \sqrt{|\omega|}$ for brevity. Then, the first feature implies

$$u_j(x) = \eta\sqrt{\frac{2k_j^2}{2k_j^2-1}}\operatorname{cn}\left(\frac{\eta}{\sqrt{2k_j^2-1}}\left(x+a_j\right); k_j\right), \qquad k_j \in \left(\frac{1}{\sqrt{2}}, 1\right),\ a_j \in \left[0, T_j\right),\ j = 1,2 \tag{8}$$

where $\operatorname{cn}(\cdot; k)$ is the cnoidal function of parameter k and $T_j = T_j\left(k_j, \eta\right) := S\left(k_j\right)/\eta$ is the period of the function $\operatorname{cn}\left(\eta\left(\cdot\right)/\sqrt{2k_j^2-1}; k_j\right)$. Here and in the rest of the paper, S denotes the function

$$S(k) := 4\sqrt{2k^2-1}\,K(k) = 4\sqrt{2k^2-1}\int_0^1 \frac{dt}{\sqrt{(1-t^2)(1-k^2t^2)}}, \tag{9}$$

where $K(k)$ is the so called complete elliptic integral of first kind. Notice that $S\colon (1/\sqrt{2}, 1) \to \mathbb{R}$ is strictly increasing, continuous and such that $S\left((1/\sqrt{2}, 1)\right) = (0, +\infty)$.

Therefore, restricting ourselves for simplicity to the case with u_3 and u_4 of the same sign, which we may assume positive thanks to the odd parity of the system in Equation (5), we are led to study the existence of solutions $\eta > 0$, $k_1, k_2 \in \left(\frac{1}{\sqrt{2}}, 1\right)$, $a_1 \in [0, T_1)$, $a_2 \in [0, T_2)$, $a_3, a_4 \in \mathbb{R}$ to the following system:

$$\begin{cases} \frac{k_1}{\sqrt{2k_1^2-1}}\,\mathrm{cn}\left(\frac{\eta a_1}{\sqrt{2k_1^2-1}};k_1\right) = \frac{k_2}{\sqrt{2k_2^2-1}}\,\mathrm{cn}\left(\frac{\eta(L+a_2)}{\sqrt{2k_2^2-1}};k_2\right) = \mathrm{sech}\,(\eta a_3) \\ \frac{k_1}{\sqrt{2k_1^2-1}}\,\mathrm{cn}\left(\frac{\eta(L_1+a_1)}{\sqrt{2k_1^2-1}};k_1\right) = \frac{k_2}{\sqrt{2k_2^2-1}}\,\mathrm{cn}\left(\frac{\eta(L_1+a_2)}{\sqrt{2k_2^2-1}};k_2\right) = \mathrm{sech}\,(\eta a_4) \\ \tanh(\eta a_3)\,\mathrm{sech}\,(\eta a_3) = \\ \quad = -\frac{k_1}{2k_1^2-1}\,\mathrm{sn}\left(\frac{\eta a_1}{\sqrt{2k_1^2-1}};k_1\right)\mathrm{dn}\left(\frac{\eta a_1}{\sqrt{2k_1^2-1}};k_1\right) + \frac{k_2}{2k_2^2-1}\,\mathrm{sn}\left(\frac{\eta(L+a_2)}{\sqrt{2k_2^2-1}};k_2\right)\mathrm{dn}\left(\frac{\eta(L+a_2)}{\sqrt{2k_2^2-1}};k_2\right) \\ \tanh(\eta a_4)\,\mathrm{sech}\,(\eta a_4) = \\ \quad = \frac{k_1}{2k_1^2-1}\,\mathrm{sn}\left(\frac{\eta(L_1+a_1)}{\sqrt{2k_1^2-1}};k_1\right)\mathrm{dn}\left(\frac{\eta(L_1+a_1)}{\sqrt{2k_1^2-1}};k_1\right) - \frac{k_2}{2k_2^2-1}\,\mathrm{sn}\left(\frac{\eta(L_1+a_2)}{\sqrt{2k_2^2-1}};k_2\right)\mathrm{dn}\left(\frac{\eta(L_1+a_2)}{\sqrt{2k_2^2-1}};k_2\right). \end{cases} \tag{10}$$

This set of equations turns out to be still rather difficult to study in his full generality, and indeed we have results only in the subcase where the two solitons in Equation (7) have the same height at the vertices, i.e., $\mathrm{sech}\,(\eta a_3) = \mathrm{sech}\,(\eta a_4)$ (which corresponds to $\theta_1 = \theta_2$ in Section 2). More precisely, in Section 2 we reduce the system in Equation (10) to an equivalent one, which naturally splits into different cases. Then, we study three of such cases, all with $\mathrm{sech}\,(\eta a_3) = \mathrm{sech}\,(\eta a_4)$, leading to our existence results, which are the following three theorems.

The first two results only concern the case of irrational ratios L_1/L_2 and give solutions with $k_1 \neq k_2$, i.e., non-periodic on the ring of the graph.

Theorem 1. *Assume that $L_1/L_2 \in \mathbb{R} \setminus \mathbb{Q}$. Then, there exists a sequence of positive integers $(n_h)_{h\in\mathbb{N}}$ such that for every $\omega < -32K(1/\sqrt{2})^2/(L_1L_2)$ there exists $h_\omega \in \mathbb{N}$ (also depending on L_1 and L_2) such that for all $h > h_\omega$ the problem in Equation (5) has two solutions $(u^+_{1,h}, u^+_{2,h}, u^+_{3,h}, u^+_{4,h})$ and $(u^-_{1,h}, u^-_{2,h}, u^-_{3,h}, u^-_{4,h})$ of the form:*

$$u^{\pm}_{j,h}(x) = \sqrt{\frac{2\,|\omega|\,k^2_{j,h}}{2k^2_{j,h}-1}}\,\mathrm{cn}\left(\sqrt{\frac{|\omega|}{2k^2_{j,h}-1}}\left(x + a^{\pm}_{j,h}\right); k_{j,h}\right), \qquad j = 1,2 \tag{11}$$

$$u^{\pm}_{j,h}(x) = \sqrt{2\,|\omega|}\,\mathrm{sech}\left(\sqrt{|\omega|}\left(x + a^{\pm}_{j,h}\right)\right), \qquad j = 3,4 \tag{12}$$

where $u^{\pm}_{1,h}(x)$ and $u^{\pm}_{2,h}(x)$ have periods $T_{1,h} = L_1/\left[n_hL_1/L_2 + 1\right]$ and $T_{2,h} = L_2/n_h$, and for all h one has

$$\frac{1}{\sqrt{2}} < k_{1,h} < k_{2,h} < 1, \quad a^{\pm}_{1,h} \in \left(0, \frac{T_{1,h}}{4}\right), \quad a^{\pm}_{2,h} \in [0, T_{2,h}), \quad a^{\pm}_{3,h} < 0, \quad a^{\pm}_{4,h} > 0, \quad a^+_{j,h} \neq a^-_{j,h}. \tag{13}$$

Remark 1. *More precisely, according to the proof, in Theorem 1, we have that*

$$k_{1,h} = S^{-1}\left(\frac{L_1}{[n_hL_1/L_2+1]}\sqrt{|\omega|}\right), \qquad a^{\pm}_{1,h} = \gamma_1(k_{1,h}, \omega, \theta^{\pm}_h),$$
$$k_{2,h} = S^{-1}\left(\frac{L_2}{n_h}\sqrt{|\omega|}\right), \qquad a^{\pm}_{2,h} = \gamma_2(k_{2,h}, \omega, \theta^{\pm}_h) - L + pT_{2,h}, \qquad -a^{\pm}_{3,h} = a^{\pm}_{4,h} = \mathrm{sech}^{-1}_{|[0,+\infty)}(\theta^{\pm}_h),$$

where p is the unique positive integer such that $a^{\pm}_{2,h} \in [0, T_{2,h})$, $\theta^{\pm}_h$ are the two distinct solutions in $(0,1]$ of the equation $\theta^2\left(1-\theta^2\right) = t_{k_{1,h},k_{2,h}}$ with $t_{k_{1,h},k_{2,h}}$ given by Equation (17), and $\gamma_j(k_{j,h}, \omega, \theta^{\pm}_h)$ is the unique preimage in $\left(0, T_{j,h}/4\right)$ of $\theta^{\pm}_h\sqrt{2k^2_{j,h}-1}/k_{j,h}$ by the function $\mathrm{cn}\left((\cdot)\sqrt{|\omega|}/\sqrt{2k^2_{j,h}-1}; k_{j,h}\right)$.

Theorem 2. *Assume that $L_1/L_2 \in \mathbb{R} \setminus \mathbb{Q}$. Then, there exists a sequence of positive integers $(n_h)_{h\in\mathbb{N}}$ such that for every $\omega < -32K(1/\sqrt{2})^2/(L_1L_2)$ there exists $h_\omega \in \mathbb{N}$ (also depending on L_1 and L_2) such that for all $h > h_\omega$ the problem in Equation (5) has two solutions $(u^{\pm}_{1,h}, u^{\pm}_{2,h}, u^{\pm}_{3,h}, u^{\pm}_{4,h})$ of the form of Equations (11)–(12), where $u^{\pm}_{1,h}(x)$ and $u^{\pm}_{2,h}(x)$ have periods $T_{1,h} = L_1/\left[n_hL_1/L_2\right]$ and $T_{2,h} = L_2/n_h$, the parameters $a^{\pm}_{1,h}, a^{\pm}_{2,h}, a^{\pm}_{3,h}, a^{\pm}_{4,h}$ are as in Equation (13) and for all h one has*

$$\frac{1}{\sqrt{2}} < k_{2,h} < k_{1,h} < 1.$$

Remark 2. *More precisely, in Theorem 2 we have that*

$$k_{1,h} = S^{-1}\left(\frac{L_1}{[n_h L_1/L_2]}\sqrt{|\omega|}\right) \quad \text{and} \quad k_{2,h} = S^{-1}\left(\frac{L_2}{n_h}\sqrt{|\omega|}\right),$$

whereas $a^{\pm}_{j,h}$ *are exactly as in Remark 1.*

The third result does not need L_1/L_2 irrational and concerns the subcase of the system in Equation (5) which, if $L_1/L_2 \in \mathbb{R} \setminus \mathbb{Q}$ and $k_1 = k_2$, is exactly the system in Equation (6) with plus sign (see Remark 5).

Theorem 3. *Let $m, n \in \mathbb{N}$ be such that $n > m \geq 1$. Then, there exists $\omega_{m,n} < 0$ (also depending on L_1) such that for all $\omega < \omega_{m,n}$ the problem in Equation (5) has a solution (u_1, u_2, u_3, u_4) of the form of Equations (7)–(8), with $k_1, k_2 \in \left(\sqrt{3}/2, 1\right)$, $a_1 \in (0, T_1/4)$, $a_2 \in [0, T_2)$.*

Remark 3. *According to the proof, in Theorem 3, a_1, a_2, a_3, a_4 can be described in a similar way of Theorems 1 and 2. On the contrary, the parameters k_1, k_2 do exist, but are not explicit as in the previous theorems.*

As already mentioned, Theorems 1–3 do not exhaust the study of solutions to the problem in Equation (5), and thus of standing waves of (NLS), as they only concern the case of solitons having the same height at the vertices. In addition, they do not describe the whole family of this kind of solutions, but only give existence results. However, they still provide some answer to the questions raised above. Indeed, Theorems 1 and 2 answer in the affirmative to the first question, as they prove existence of standing waves which are non-periodic on the ring of $\mathcal{G}$. As to Theorem 3, for any m and n, it provides a family of solutions which depend on the continuous parameter $\omega \in (-\infty, \omega_{m,n})$ and, roughly speaking, make m oscillations on the edge of length L_1 and $n - m$ oscillations on the one of length L_2 (cf. the second and third equations of the system in Equation (33)). If L_1/L_2 is irrational and one of these families contain a solution with $k_1 = k_2$, then such a solution is one of the isolated solutions found in [15] in the irrational case and we can answer affirmatively also to the second question. Unfortunately, the argument we used in proving Theorem 3 does not allow us to say wether we find solutions with $k_1 = k_2$ or not, and therefore we do not have a final answer to the second question.

2. Preliminaries

In this section, we reduce the system in Equation (10) to a simpler equivalent one, which is the system in Equation (14) with the last two equations replaced by the system in Equation (19).

For brevity, we set

$$X_1 = \frac{\eta a_1}{\sqrt{2k_1^2 - 1}}, \quad X_2 = \frac{\eta (L + a_2)}{\sqrt{2k_2^2 - 1}}, \quad X_3 = \frac{\eta (L_1 + a_1)}{\sqrt{2k_1^2 - 1}}, \quad X_4 = \frac{\eta (L_1 + a_2)}{\sqrt{2k_2^2 - 1}},$$

and

$$\sigma_1 = \operatorname{sgn}[\operatorname{sn}(X_1; k_1)], \quad \sigma_2 = \operatorname{sgn}[\operatorname{sn}(X_2; k_2)], \quad \sigma_3 = \operatorname{sgn}[\operatorname{sn}(X_3; k_1)], \quad \sigma_4 = \operatorname{sgn}[\operatorname{sn}(X_4; k_2)].$$

Then, using well known identities (see [20]) and the first equation of the system in Equation (10), we get

$$\begin{aligned}\operatorname{sn}(X_1;k_1) &= \sigma_1\sqrt{1-\operatorname{cn}^2(X_1;k_1)} = \sigma_1\sqrt{1-\frac{2k_1^2-1}{k_1^2}\operatorname{sech}^2(\eta a_3)},\\ \operatorname{dn}(X_1;k_1) &= \sqrt{1-k_1^2+k_1^2\operatorname{cn}^2(X_1;k_1)} = \sqrt{1-k_1^2+(2k_1^2-1)\operatorname{sech}^2(\eta a_3)}\end{aligned}$$

and hence

$$\begin{aligned}\frac{k_1}{2k_1^2-1}\operatorname{sn}(X_1;k_1)\operatorname{dn}(X_1;k_1) &= \sigma_1\sqrt{\frac{k_1^2}{2k_1^2-1}-\operatorname{sech}^2(\eta a_3)}\sqrt{\frac{(1-k_1^2)}{2k_1^2-1}+\operatorname{sech}^2(\eta a_3)}\\ &= \sigma_1\sqrt{\frac{k_1^2(1-k_1^2)}{(2k_1^2-1)^2}+\operatorname{sech}^2(\eta a_3)-\operatorname{sech}^4(\eta a_3)}.\end{aligned}$$

Arguing similarly for the products $\operatorname{sn}(X_2;k_2)\operatorname{dn}(X_2;k_2)$, $\operatorname{sn}(X_3;k_1)\operatorname{dn}(X_3;k_1)$ and $\operatorname{sn}(X_4;k_2)\operatorname{dn}(X_4;k_2)$, and defining

$$c(k) := \frac{k^2(1-k^2)}{(2k^2-1)^2},$$

we thus obtain that the system in Equation (10) is equivalent to

$$\begin{cases}\frac{k_1}{\sqrt{2k_1^2-1}}\operatorname{cn}\left(\frac{\eta a_1}{\sqrt{2k_1^2-1}};k_1\right) = \frac{k_2}{\sqrt{2k_2^2-1}}\operatorname{cn}\left(\frac{\eta(L+a_2)}{\sqrt{2k_2^2-1}};k_2\right) = \operatorname{sech}(\eta a_3)\\ \frac{k_1}{\sqrt{2k_1^2-1}}\operatorname{cn}\left(\frac{\eta(L_1+a_1)}{\sqrt{2k_1^2-1}};k_1\right) = \frac{k_2}{\sqrt{2k_2^2-1}}\operatorname{cn}\left(\frac{\eta(L_1+a_2)}{\sqrt{2k_2^2-1}};k_2\right) = \operatorname{sech}(\eta a_4)\\ \tanh(\eta a_3)\operatorname{sech}(\eta a_3) = -\sigma_1\sqrt{c(k_1)+\operatorname{sech}^2(\eta a_3)-\operatorname{sech}^4(\eta a_3)}+\sigma_2\sqrt{c(k_2)+\operatorname{sech}^2(\eta a_3)-\operatorname{sech}^4(\eta a_3)}\\ \tanh(\eta a_4)\operatorname{sech}(\eta a_4) = \sigma_3\sqrt{c(k_1)+\operatorname{sech}^2(\eta a_4)-\operatorname{sech}^4(\eta a_4)}-\sigma_4\sqrt{c(k_2)+\operatorname{sech}^2(\eta a_4)-\operatorname{sech}^4(\eta a_4)}.\end{cases} \tag{14}$$

Let us now focus on the last two equations. Setting

$$\theta_1 = \operatorname{sech}(\eta a_3),\ \theta_2 = \operatorname{sech}(\eta a_4),\ \sigma_5 = \operatorname{sgn}(a_3) = \operatorname{sgn}(\tanh(\eta a_3)),\ \sigma_6 = \operatorname{sgn}(a_4) = \operatorname{sgn}(\tanh(\eta a_4))$$

the couple of such equations is equivalent to

$$\begin{cases}\sigma_5\sqrt{1-\theta_1^2}\,\theta_1 = -\sigma_1\sqrt{c(k_1)+\theta_1^2(1-\theta_1^2)}+\sigma_2\sqrt{c(k_2)+\theta_1^2(1-\theta_1^2)}\\ \sigma_6\sqrt{1-\theta_2^2}\,\theta_2 = \sigma_3\sqrt{c(k_1)+\theta_2^2(1-\theta_2^2)}-\sigma_4\sqrt{c(k_2)+\theta_2^2(1-\theta_2^2)}.\end{cases} \tag{15}$$

Squaring the equations, we get

$$c(k_1)+\theta_1^2\left(1-\theta_1^2\right)+c(k_2)-2\sigma_1\sigma_2\sqrt{c(k_1)+\theta_1^2(1-\theta_1^2)}\sqrt{c(k_2)+\theta_1^2(1-\theta_1^2)} = 0,$$

$$c(k_1)+\theta_2^2\left(1-\theta_2^2\right)+c(k_2)-2\sigma_3\sigma_4\sqrt{c(k_1)+\theta_2^2(1-\theta_2^2)}\sqrt{c(k_2)+\theta_2^2(1-\theta_2^2)} = 0,$$

which are impossible if $\sigma_1\sigma_2 = -1$ or $\sigma_3\sigma_4 = -1$. Hence, we can add the conditions $\sigma_1 = \sigma_2$ and $\sigma_3 = \sigma_4$ to the system in Equation (15), and get

$$\begin{cases}\sigma_5\sqrt{1-\theta_1^2}\,\theta_1 = \sigma_1\left(-\sqrt{c(k_1)+\theta_1^2(1-\theta_1^2)}+\sqrt{c(k_2)+\theta_1^2(1-\theta_1^2)}\right)\\ \sigma_6\sqrt{1-\theta_2^2}\,\theta_2 = \sigma_3\left(\sqrt{c(k_1)+\theta_2^2(1-\theta_2^2)}-\sqrt{c(k_2)+\theta_2^2(1-\theta_2^2)}\right)\\ \sigma_2 = \sigma_1,\quad \sigma_4 = \sigma_3.\end{cases} \tag{16}$$

Moreover, both $\theta_1^2\left(1-\theta_1^2\right)$ and $\theta_2^2\left(1-\theta_2^2\right)$ must be solutions $t \in [0,1/4]$ of the equation

$$c\left(k_1\right)+c\left(k_2\right)+t-2\sqrt{c\left(k_1\right)+t}\sqrt{c\left(k_2\right)+t}=0.$$

Such equation has the unique nonnegative solution

$$t=t_{k_1,k_2}=\frac{1}{3}\left(2\sqrt{c\left(k_1\right)^2-c\left(k_1\right)c\left(k_2\right)+c\left(k_2\right)^2}-c\left(k_1\right)-c\left(k_2\right)\right), \tag{17}$$

which belongs to $[0,1/4]$ if and only if (k_1,k_2) belongs to the set

$$A=\left\{(k_1,k_2)\in\left(\frac{1}{\sqrt{2}},1\right)^2 : 2\sqrt{c\left(k_1\right)^2-c\left(k_1\right)c\left(k_2\right)+c\left(k_2\right)^2}-c\left(k_1\right)-c\left(k_2\right)\leq\frac{3}{4}\right\},$$

i.e., as one can easily see after some computations,

$$A=\left\{(k_1,k_2)\in\mathbb{R}: \quad k_1\in\left(\frac{1}{\sqrt{2}},1\right), \quad \frac{\sqrt{4k_1^2-1}}{2k_1}\leq k_2\leq\frac{1}{2\sqrt{1-k_1^2}}, \quad k_2<1\right\}$$

(the set A is portrayed in Figure 4).

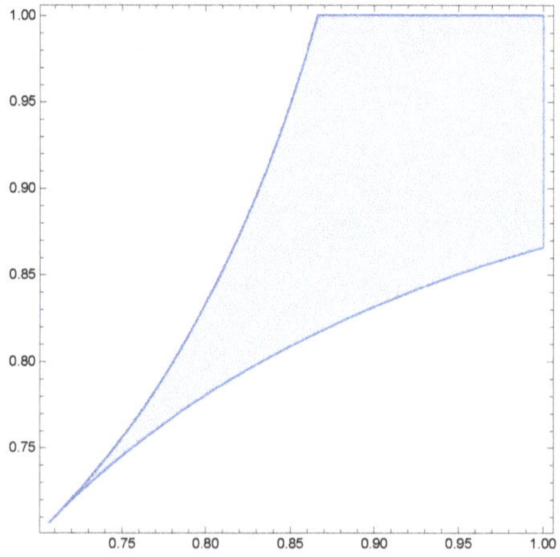

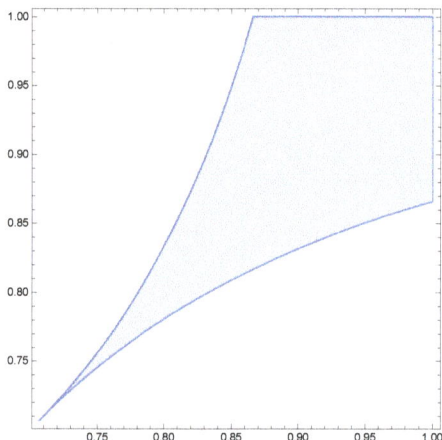

Figure 4. The set A. The point $(\sqrt{2}/2, \sqrt{2}/2)$ and the straight lines of the boundary are not included.

In this case, the equation $\theta^2\left(1-\theta^2\right)=t_{k_1,k_2}$ with $\theta\in(0,1]$ has two distinct solutions

$$\theta_{k_1,k_2}^{\pm}=\sqrt{\frac{1\pm\sqrt{1-4t_{k_1,k_2}}}{2}} \tag{18}$$

if $t_{k_1,k_2}\in(0,1/4)$, two coincident solutions $\theta_{k_1,k_2}^{+}=\theta_{k_1,k_2}^{-}=1/\sqrt{2}$ if $t_{k_1,k_2}=1/4$, and a unique solution $\theta_{k_1,k_2}^{+}=1$ if $t_{k_1,k_2}=0$ (i.e., $k_1=k_2$). In this latter case, we still write $\theta_{k_1,k_2}^{+}=\theta_{k_1,k_2}^{-}=1$ for future convenience. We also observe that the function $c(k)$ is positive and strictly decreasing from $\left(1/\sqrt{2},1\right)$ onto $(0,+\infty)$, so that the terms within brackets on the right hand sides of the first two equations of Equation (16) have a fixed sign according as $k_1<k_2$ or $k_1>k_2$. Therefore, the system in Equation (15) turns out to be equivalent to

$$\begin{cases} (k_1,k_2) \in A, \quad \theta_1,\theta_2 \in \left\{\theta^+_{k_1,k_2},\theta^-_{k_1,k_2}\right\} \\ \operatorname{sech}(\eta a_3) = \theta_1, \quad \operatorname{sech}(\eta a_4) = \theta_2 \\ \begin{cases} k_1 < k_2 \\ \sigma_5 = -\sigma_1 \\ \sigma_6 = \sigma_3 \end{cases} \vee \begin{cases} k_1 > k_2 \\ \sigma_5 = \sigma_1 \\ \sigma_6 = -\sigma_3 \end{cases} \vee \begin{cases} k_1 = k_2 \\ a_3 = a_4 = 0 \end{cases} \\ \sigma_2 = \sigma_1, \quad \sigma_4 = \sigma_3. \end{cases} \tag{19}$$

As a conclusion, Equation (10) is equivalent to the system in Equation (14) with the last two equations replaced by the system in Equation (19).

3. Case $\theta_1 = \theta_2$, $\sigma_1 = \sigma_3$ and $k_1 < k_2$. Proof of Theorem 1

We focus on the case $\sigma_1 = \sigma_3 = 1$, which gives Theorem 1, leaving the analogous case $\sigma_1 = \sigma_3 = -1$ to the interested reader. In such a case, condition $(k_1,k_2) \in A$ becomes

$$(k_1,k_2) \in A' = A \cap \{(k_1,k_2) \in \mathbb{R} : k_1 < k_2\} = \left\{(k_1,k_2) \in \mathbb{R} : \quad \frac{1}{\sqrt{2}} < k_1 < k_2 \leq \frac{1}{2\sqrt{1-k_1^2}}, \quad k_2 < 1\right\}$$

and, taking into account the equivalence between Equation (15) and Equation (19), the system in Equation (14) becomes:

$$\begin{cases} (k_1,k_2) \in A', \quad \theta \in \left\{\theta^+_{k_1,k_2},\theta^-_{k_1,k_2}\right\} \\ \operatorname{sech}(\eta a_3) = \operatorname{sech}(\eta a_4) = \theta, \quad a_3 < 0, \quad a_4 > 0 \\ \frac{k_1}{\sqrt{2k_1^2-1}} \operatorname{cn}\left(\frac{\eta a_1}{\sqrt{2k_1^2-1}}; k_1\right) = \frac{k_2}{\sqrt{2k_2^2-1}} \operatorname{cn}\left(\frac{\eta(L+a_2)}{\sqrt{2k_2^2-1}}; k_2\right) = \theta \\ \frac{k_1}{\sqrt{2k_1^2-1}} \operatorname{cn}\left(\frac{\eta(L_1+a_1)}{\sqrt{2k_1^2-1}}; k_1\right) = \frac{k_2}{\sqrt{2k_2^2-1}} \operatorname{cn}\left(\frac{\eta(L_1+a_2)}{\sqrt{2k_2^2-1}}; k_2\right) = \theta \\ \operatorname{sn}\left(\frac{\eta a_1}{\sqrt{2k_1^2-1}}; k_1\right) > 0, \quad \operatorname{sn}\left(\frac{\eta(L_1+a_1)}{\sqrt{2k_1^2-1}}; k_1\right) > 0 \\ \operatorname{sn}\left(\frac{\eta(L+a_2)}{\sqrt{2k_2^2-1}}; k_2\right) > 0, \quad \operatorname{sn}\left(\frac{\eta(L_1+a_2)}{\sqrt{2k_2^2-1}}; k_2\right) > 0. \end{cases} \tag{20}$$

We denote by $\gamma_j = \gamma_j(k_j,\eta,\theta)$ the unique preimage in $(0,T_j/4)$ of the value $\frac{\sqrt{2k_j^2-1}}{k_j}\theta$ by the function $\operatorname{cn}\left(\frac{\eta}{\sqrt{2k_j^2-1}}(\cdot); k_j\right)$. Then,

$$\begin{cases} \frac{k_1}{\sqrt{2k_1^2-1}} \operatorname{cn}\left(\frac{\eta a_1}{\sqrt{2k_1^2-1}}; k_1\right) = \theta, \quad \operatorname{sn}\left(\frac{\eta a_1}{\sqrt{2k_1^2-1}}; k_1\right) > 0 \\ \frac{k_1}{\sqrt{2k_1^2-1}} \operatorname{cn}\left(\frac{\eta(L_1+a_1)}{\sqrt{2k_1^2-1}}; k_1\right) = \theta, \quad \operatorname{sn}\left(\frac{\eta(L_1+a_1)}{\sqrt{2k_1^2-1}}; k_1\right) > 0 \end{cases}$$

means

$$\begin{cases} a_1 = \gamma_1 \\ L_1 + a_1 = \gamma_1 + mT_1 \quad \text{for some } m \geq 1, \end{cases} \qquad \text{i.e.,} \qquad \begin{cases} a_1 = \gamma_1 \\ L_1 = mT_1 \quad \text{for some } m \geq 1 \end{cases}$$

while

$$\begin{cases} \frac{k_2}{\sqrt{2k_2^2-1}} \operatorname{cn}\left(\frac{\eta(L+a_2)}{\sqrt{2k_2^2-1}};k_2\right) = \theta, & \operatorname{sn}\left(\frac{\eta(L+a_2)}{\sqrt{2k_2^2-1}};k_2\right) > 0 \\ \frac{k_2}{\sqrt{2k_2^2-1}} \operatorname{cn}\left(\frac{\eta(L_1+a_2)}{\sqrt{2k_2^2-1}};k_2\right) = \theta, & \operatorname{sn}\left(\frac{\eta(L_1+a_2)}{\sqrt{2k_2^2-1}};k_2\right) > 0 \end{cases}$$

means

$$\begin{cases} L + a_2 = \gamma_2 + pT_2 & \text{for some } p \geq 0 \\ L_1 + a_2 = \gamma_2 + qT_2 & \text{for some } 0 \leq q < p, \end{cases} \quad \text{i.e.,} \quad \begin{cases} L + a_2 = \gamma_2 + pT_2 & \text{for some } p \geq 0 \\ L_2 = (p-q)\,T_2 & \text{for some } 0 \leq q < p. \end{cases}$$

Hence, the system in Equation (20) becomes

$$\begin{cases} (k_1, k_2) \in A', \quad \theta \in \left\{\theta^+_{k_1,k_2}, \theta^-_{k_1,k_2}\right\} \\ \operatorname{sech}(\eta a_3) = \operatorname{sech}(\eta a_4) = \theta, \quad a_3 < 0, \quad a_4 > 0 \\ L_1 = mT_1(k_1,\eta) \quad \text{for some } m \geq 1 \\ L_2 = nT_2(k_2,\eta) \quad \text{for some } n \geq 1 \\ a_1 = \gamma_1(k_1,\eta,\theta) \\ a_2 = \gamma_2(k_2,\eta,\theta) + pT_2(k_2,\eta) - L \quad \text{for some } p \geq n \end{cases} \tag{21}$$

(observe that θ depends on both k_1 and k_2, and so do a_1 and a_2 according to the last two equations).

Remark 4. *The equivalence between the systems in Equation (20) and Equation (21) does not need assumption $k_1 < k_2$. On the other hand, if $k_1 = k_2$, then $T_1(k_1,\eta) = T_2(k_2,\eta)$ and thus the third and fourth equations of the system in Equation (21) imply $L_1/L_2 \in \mathbb{Q}$. This means that solutions to the system in Equation (10) with $k_1 = k_2$ (which implies $\theta_1 = \theta_2 = 1$) and $\sigma_1 = \sigma_3$ cannot exist if the ratio L_1/L_2 is not rational.*

Let us now focus on the following group of equations:

$$\begin{cases} (k_1, k_2) \in A' \\ L_1 = mT_1(k_1,\eta), \quad \text{for some } m \geq 1 \\ L_2 = nT_2(k_2,\eta), \quad \text{for some } n \geq 1. \end{cases} \tag{22}$$

Recalling that $T_j(k_j,\eta) = S(k_j)/\eta$, this system is equivalent to

$$\begin{cases} \frac{1}{\sqrt{2}} < k_1 < k_2 \leq \frac{1}{2\sqrt{1-k_1^2}}, \; k_2 < 1 \\ k_1 = S^{-1}\left(\eta\frac{L_1}{m}\right) \quad \text{for some } m \geq 1 \\ k_2 = S^{-1}\left(\eta\frac{L_2}{n}\right) \quad \text{for some } n \geq 1. \end{cases} \tag{23}$$

and therefore, recalling that S is strictly increasing and continuous from $(1/\sqrt{2}, 1)$ onto $(0, +\infty)$, we can obtain solutions by fixing $\eta > 0$ and finding $n, m \geq 1$ such that

$$\begin{cases} S^{-1}\left(\eta\frac{L_1}{m}\right) < S^{-1}\left(\eta\frac{L_2}{n}\right) \\ S^{-1}\left(\eta\frac{L_2}{n}\right) \leq \frac{1}{2\sqrt{1-\left[S^{-1}\left(\eta\frac{L_1}{m}\right)\right]^2}}, \end{cases} \quad \text{i.e.,} \quad \begin{cases} \frac{L_1}{m} < \frac{L_2}{n} \\ \eta\frac{L_2}{n} \leq S\left(\frac{1}{2\sqrt{1-\left[S^{-1}\left(\eta\frac{L_1}{m}\right)\right]^2}}\right). \end{cases} \tag{24}$$

Lemma 1. *One has*

$$S\left(\frac{1}{2\sqrt{1-\left[S^{-1}(t)\right]^2}}\right)=t+\frac{1}{32K_0^2}t^3+o\left(t^3\right) \qquad \text{as } t\to 0^+$$

(where, we recall, $K_0=K\left(1/\sqrt{2}\right)$*).*

Proof. We have

$$\begin{aligned}
\lim_{t\to 0^+}\frac{S^{-1}(t)-\frac{1}{\sqrt{2}}-\frac{t^2}{32K_0^2\sqrt{2}}}{t^4} &= \lim_{k\to(1/\sqrt{2})^+}\frac{S^{-1}(S(k))-\frac{1}{\sqrt{2}}-\frac{S(k)^2}{32K_0^2\sqrt{2}}}{S(k)^4}\\
&= \lim_{k\to(1/\sqrt{2})^+}\frac{k-\frac{1}{\sqrt{2}}-\frac{16K(k)^2(2k^2-1)}{32K_0^2\sqrt{2}}}{2^8K(k)^4(2k^2-1)^2}\\
&= \frac{1}{2^{10}K_0^2}\lim_{k\to(1/\sqrt{2})^+}\frac{2K_0^2-K(k)^2\left(\sqrt{2}k+1\right)}{K(k)^4\left(\sqrt{2}k+1\right)^2\left(k-1/\sqrt{2}\right)}
\end{aligned}$$

where, setting $K_0'=K'\left(1/\sqrt{2}\right)$, by De L'Hôpital's rule, we get

$$\lim_{k\to(1/\sqrt{2})^+}\frac{2K_0^2-K(k)^2\left(\sqrt{2}k+1\right)}{k-1/\sqrt{2}}=-4K_0K_0'-K_0^2\sqrt{2}.$$

Hence, we conclude

$$\lim_{t\to 0^+}\frac{S^{-1}(t)-\frac{1}{\sqrt{2}}-\frac{t^2}{32K_0^2\sqrt{2}}}{t^4}=-\frac{K_0+2\sqrt{2}K_0'}{2^{11}\sqrt{2}K_0^5},$$

i.e.,

$$S^{-1}(t)=\frac{1}{\sqrt{2}}+c_1t^2-c_2t^4+o\left(t^4\right) \qquad \text{as } t\to 0^+ \tag{25}$$

where $c_1=\frac{1}{32\sqrt{2}K_0^2}$ and $c_2=\frac{K_0+2\sqrt{2}K_0'}{2^{11}\sqrt{2}K_0^5}$. This implies

$$\begin{aligned}
\frac{1}{2\sqrt{1-S^{-1}(t)^2}} &= \frac{1}{2\sqrt{\frac{1}{2}-\frac{2}{\sqrt{2}}c_1t^2-\left(c_1^2-\sqrt{2}c_2\right)t^4+o\left(t^4\right)}}\\
&= \frac{1}{\sqrt{2}\sqrt{1-2\sqrt{2}c_1t^2-2\left(c_1^2-\sqrt{2}c_2\right)t^4+o\left(t^4\right)}}\\
&= \frac{1}{\sqrt{2}}+c_1t^2+\left(2\sqrt{2}c_1^2-c_2\right)t^4+o\left(t^4\right).
\end{aligned}$$

Using De L'Hôpital's rule again, we now compute

$$\lim_{k\to(1/\sqrt{2})^+}\frac{S(k)-2^{11/4}K_0\left(k-1/\sqrt{2}\right)^{1/2}}{\left(k-1/\sqrt{2}\right)^{3/2}}=\lim_{k\to(1/\sqrt{2})^+}\frac{S'(k)-2^{7/4}K_0\left(k-1/\sqrt{2}\right)^{-1/2}}{\frac{3}{2}\left(k-1/\sqrt{2}\right)^{1/2}}$$

$$
\begin{aligned}
&= \frac{2}{3}\lim_{k\to(1/\sqrt{2})^+}\frac{\frac{8k}{\sqrt{2k^2-1}}K(k)+4\sqrt{2k^2-1}K'(k)-\frac{2^{7/4}K_0}{\left(k-1/\sqrt{2}\right)^{1/2}}}{\left(k-1/\sqrt{2}\right)^{1/2}}\\
&= \frac{2^{15/4}}{3}K_0'+\frac{2}{3}\lim_{k\to(1/\sqrt{2})^+}\frac{\frac{8kK(k)}{\sqrt[4]{2}\sqrt{\sqrt{2}k+1}}-2^{7/4}K_0}{k-1/\sqrt{2}}\\
&= \frac{2^{15/4}}{3}K_0'+\frac{2}{3}\lim_{k\to(1/\sqrt{2})^+}\frac{\frac{8k(K(k)-K_0)}{\sqrt[4]{2}\sqrt{\sqrt{2}k+1}}+\left(\frac{8k}{\sqrt[4]{2}\sqrt{\sqrt{2}k+1}}-2^{7/4}\right)K_0}{k-1/\sqrt{2}}=2^{5/4}K_0+2^{11/4}K_0'
\end{aligned}
$$

where the result follows because $K(k)-K_0\sim K_0'\left(k-1/\sqrt{2}\right)$ as $k\to\left(1/\sqrt{2}\right)^+$ and

$$
\begin{aligned}
\frac{8k}{\sqrt[4]{2}\sqrt{\sqrt{2}k+1}}-2^{7/4} &= 2^{7/4}\frac{2k-\sqrt{\sqrt{2}k+1}}{\sqrt{\sqrt{2}k+1}}=2^{7/4}\frac{4k^2-\sqrt{2}k-1}{\sqrt{\sqrt{2}k+1}\left(2k+\sqrt{\sqrt{2}k+1}\right)}\\
&= 2^{7/4}\frac{\left(4k+\sqrt{2}\right)\left(k-1/\sqrt{2}\right)}{\sqrt{\sqrt{2}k+1}\left(2k+\sqrt{\sqrt{2}k+1}\right)}.
\end{aligned}
$$

This means

$$
S(k)=2^{11/4}K_0\left(k-1/\sqrt{2}\right)^{1/2}+\left(2^{5/4}K_0+2^{11/4}K_0'\right)\left(k-1/\sqrt{2}\right)^{3/2}+o\left(\left(k-1/\sqrt{2}\right)^{3/2}\right) \tag{26}
$$

as $k\to\left(1/\sqrt{2}\right)^+$ and therefore we deduce that as $t\to0^+$ one has (note that $2^{11/4}K_0\sqrt{c_1}=1$)

$$
\begin{aligned}
S\left(\frac{1}{2\sqrt{1-S^{-1}(t)^2}}\right) &= 2^{11/4}K_0\sqrt{c_1}t\left(1+\frac{2\sqrt{2}c_1^2-c_2}{c_1}t^2+o\left(t^2\right)\right)^{1/2}+\\
&\quad+\left(2^{5/4}K_0+2^{11/4}K_0'\right)c_1\sqrt{c_1}t^3\left(1+\frac{2\sqrt{2}c_1^2-c_2}{c_1}t^2+o\left(t^2\right)\right)^{3/2}+o\left(t^3\right)\\
&= t\left(1+\frac{1}{2}\frac{2\sqrt{2}c_1^2-c_2}{c_1}t^2+o\left(t^2\right)\right)+\\
&\quad+\left(2^{5/4}K_0+2^{11/4}K_0'\right)c_1\sqrt{c_1}t^3\left(1+\frac{3}{2}\frac{2\sqrt{2}c_1^2-c_2}{c_1}t^2+o\left(t^2\right)\right)+o\left(t^3\right)\\
&= t+\left(2^{11/4}K_0\sqrt{c_1}\frac{1}{2}\frac{2\sqrt{2}c_1^2-c_2}{c_1}+\left(2^{5/4}K_0+2^{11/4}K_0'\right)c_1\sqrt{c_1}\right)t^3+o\left(t^3\right).
\end{aligned}
$$

Simplifying the coefficient of t^3, this gives the result. □

Thanks to Lemma 1, the system in Equation (24) becomes

$$
0<\frac{m}{n}-\frac{L_1}{L_2}\leq\frac{L_1^3\eta^2}{32K_0^2L_2}\frac{1}{m^2}+\zeta_m \tag{27}
$$

where $(\zeta_m)_m$ is a suitable sequence (also dependent on L_1, L_2, η) such that $\zeta_m=o\left(m^{-2}\right)$ as $m\to\infty$. Notice that, according to systems (23) and (24), the equality sign in the second inequality amounts to $k_2=\frac{1}{2\sqrt{1-k_1^2}}$.

Proof of Theorem 1. Since $L_1/L_2 \in \mathbb{R} \setminus \mathbb{Q}$, by ([21], Corollary 1.9) there exist infinitely many rational numbers m/n such that

$$0 < \frac{m}{n} - \frac{L_1}{L_2} < \frac{1}{n^2}. \tag{28}$$

This implies $nL_1/L_2 < m < nL_1/L_2 + 1$ and thus $m = [nL_1/L_2 + 1]$. Since the denominators of such rationals m/n must be infinite, we may arrange them in a diverging sequence $(n_h) \subset \mathbb{N}$; accordingly, the corresponding numerators are $m_h = [n_h L_1/L_2 + 1]$. Now, let $\eta > 4\sqrt{2}K_0 (L_1 L_2)^{-1/2}$ and fix $\varepsilon > 0$ such that

$$\eta^2 > \left(\frac{L_1}{L_2} + \varepsilon\right)^2 \frac{32K_0^2 L_2}{L_1^3}.$$

Since Equation (28) implies that $m_h/n_h \to L_1/L_2$ as $h \to \infty$, for h large enough, we have that $m_h/n_h < L_1/L_2 + \varepsilon$, so that

$$\frac{1}{n_h^2} < \left(\frac{L_1}{L_2} + \varepsilon\right)^2 \frac{1}{m_h^2} < \frac{L_1^3 \eta^2}{32K_0^2 L_2} \frac{1}{m_h^2}.$$

Hence, up to further enlarging h, Equation (28) gives

$$0 < \frac{m_h}{n_h} - \frac{L_1}{L_2} < \left(\frac{L_1}{L_2} + \varepsilon\right)^2 \frac{1}{m_h^2} < \frac{L_1^3 \eta^2}{32K_0^2 L_2} \frac{1}{m_h^2} + \zeta_{m_h}, \tag{29}$$

so that n_h and m_h satisfy Equation (27). For every h, this provides solutions to the system in Equation (22) by taking $k_1 = k_{1,h} = S^{-1}(\eta L_1/m_h)$ and $k_2 = k_{2,h} = S^{-1}(\eta L_2/n_h)$, and thus solutions to the system in Equation (21) by choosing $\theta = \theta_h \in \{\theta^+_{k_{1,h},k_{2,h}}, \theta^-_{k_{1,h},k_{2,h}}\}$, taking p as the unique integer such that

$$0 \le \gamma_2(k_{2,h}, \eta, \theta_h) + pT_2(k_{2,h}, \eta) - L < T_2(k_{2,h}, \eta)$$

(where $T_2(k_{2,h}, \eta) = L_2/n_h$), which turns out to be greater than or equal to n_h, and defining a_1, a_2, a_3, a_4 according to the second, fifth and sixth equation of the system. Note that $\theta^+_{k_{1,h},k_{2,h}}$ and $\theta^-_{k_{1,h},k_{2,h}}$ are different for all h, since $t_{k_{1,h},k_{2,h}} \neq 0$ (because $k_{1,h} \neq k_{2,h}$) and $t_{k_{1,h},k_{2,h}} \neq 1/4$ (because of the strict inequality signs in Equation (29)). Up to discarding a finite number of terms of the sequence (n_h), the proof is complete. □

4. Case $\theta_1 = \theta_2$, $\sigma_1 = \sigma_3$ and $k_1 > k_2$. Proof of Theorem 2

As in the previous section, we focus on the case $\sigma_1 = \sigma_3 = 1$. In this case, the system in Equation (14) becomes again the system in Equation (21), but with $(k_1, k_2) \in A'$ replaced by $(k_1, k_2) \in A''$, where

$$A'' = A \cap \{(k_1, k_2) \in \mathbb{R} : k_1 > k_2\} = \left\{(k_1, k_2) \in \mathbb{R} : \frac{\sqrt{4k_1^2 - 1}}{2k_1} \le k_2 < k_1 < 1\right\}.$$

Then, Equation (22) is now equivalent to the system

$$\begin{cases} \sqrt{1 - \frac{1}{4k_1^2}} \le k_2 < k_1 < 1 \\ k_1 = S^{-1}\left(\eta \frac{L_1}{m}\right) & \text{for some } m \ge 1 \\ k_2 = S^{-1}\left(\eta \frac{L_2}{n}\right) & \text{for some } n \ge 1, \end{cases}$$

i.e.,

$$\begin{cases} \frac{L_2}{n} < \frac{L_1}{m} \\ \eta \frac{L_2}{n} \geq S\left(\sqrt{1 - \frac{1}{4S^{-1}\left(\eta \frac{L_1}{m}\right)^2}}\right) \\ k_1 = S^{-1}\left(\eta \frac{L_1}{m}\right), \quad k_2 = S^{-1}\left(\eta \frac{L_2}{n}\right) \end{cases} \tag{30}$$

with $\eta > 0$ and $n, m \in \mathbb{N}$.

Lemma 2. *One has*

$$S\left(\sqrt{1 - \frac{1}{4S^{-1}(t)^2}}\right) = t - \frac{1}{32K_0^2} t^3 + o\left(t^3\right) \qquad \text{as } t \to 0^+$$

(where, we recall, $K_0 = K\left(1/\sqrt{2}\right)$*).*

Proof. Since $S^{-1}(t) = \frac{1}{\sqrt{2}} + c_1 t^2 - c_2 t^4 + o(t^4)$ as $t \to 0^+$ (see Equation (25)), we have

$$\begin{aligned} 1 - \frac{1}{2S^{-1}(t)^2} &= 1 - \frac{1}{2\left(S^{-1}(t) - 1/\sqrt{2} + 1/\sqrt{2}\right)^2} \\ &= 1 - \frac{1}{2} \frac{1}{\left(S^{-1}(t) - 1/\sqrt{2}\right)^2 + 1/2 + 2\left(S^{-1}(t) - 1/\sqrt{2}\right)/\sqrt{2}} \\ &= 1 - \frac{1}{2} \frac{1}{(c_1 t^2 - c_2 t^4 + o(t^4))^2 + 1/2 + 2(c_1 t^2 - c_2 t^4 + o(t^4))/\sqrt{2}} \\ &= 1 - \frac{1}{1 + 2c_1\sqrt{2}t^2 + 2\left(c_1^2 - c_2\sqrt{2}\right)t^4 + o(t^4)} \\ &= 2c_1\sqrt{2}t^2 - 2\left(3c_1^2 + c_2\sqrt{2}\right)t^4 + o\left(t^4\right) \end{aligned}$$

and therefore

$$\begin{aligned} \sqrt{1 - \frac{1}{4S^{-1}(t)^2}} &= \frac{1}{\sqrt{2}} \sqrt{1 + \left(1 - \frac{1}{2S^{-1}(t)^2}\right)} \\ &= \frac{1}{\sqrt{2}} \left(1 + \frac{1}{2}\left(1 - \frac{1}{2S^{-1}(t)^2}\right) - \frac{1}{8}\left(1 - \frac{1}{2S^{-1}(t)^2}\right)^2 + o\left(\left(1 - \frac{1}{2S^{-1}(t)^2}\right)^2\right)\right) \\ &= \frac{1}{\sqrt{2}} + c_1 t^2 - \left(2\sqrt{2}c_1^2 + c_2\right)t^4 + o\left(t^4\right). \end{aligned}$$

Hence, using the expansion in Equation (25), we deduce that

$$\begin{aligned} S\left(\sqrt{1-\frac{1}{4S^{-1}(t)^2}}\right) &= 2^{11/4}K_0\sqrt{c_1}t\left(1-\frac{2\sqrt{2}c_1^2+c_2}{c_1}t^2+o\left(t^2\right)\right)^{1/2}+ \\ &\quad+\left(2^{5/4}K_0+2^{11/4}K_0'\right)c_1\sqrt{c_1}t^3\left(1-\frac{2\sqrt{2}c_1^2+c_2}{c_1}t^2+o\left(t^2\right)\right)^{3/2}+o\left(t^3\right) \\ &= t\left(1-\frac{1}{2}\frac{2\sqrt{2}c_1^2+c_2}{c_1}t^2+o\left(t^2\right)\right)+ \\ &\quad+\left(2^{5/4}K_0+2^{11/4}K_0'\right)c_1\sqrt{c_1}t^3\left(1-\frac{3}{2}\frac{2\sqrt{2}c_1^2+c_2}{c_1}t^2+o\left(t^2\right)\right)+o\left(t^3\right) \\ &= t+\left(\left(2^{5/4}K_0+2^{11/4}K_0'\right)c_1\sqrt{c_1}-2^{11/4}K_0\sqrt{c_1}\frac{1}{2}\frac{2\sqrt{2}c_1^2+c_2}{c_1}\right)t^3+o\left(t^3\right). \end{aligned}$$

Simplifying the coefficient of t^3, the result ensues. □

By Lemma 2, the first two conditions of the system in Equation (24) become

$$0>\frac{m}{n}-\frac{L_1}{L_2}\geq-\frac{L_1^3\eta^2}{32K_0^2L_2}\frac{1}{m^2}+\zeta_m$$

where $(\zeta_m)_m$ is a suitable sequence such that $\zeta_m=o\left(m^{-2}\right)$ as $m\to\infty$. Notice that the equality sign in the second inequality amounts to $k_2=\frac{\sqrt{4k_1^2-1}}{2k_1}$.

Proof of Theorem 2. Since $L_1/L_2\in\mathbb{R}\setminus\mathbb{Q}$, by ([21], Corollary 1.9) there exist infinitely many rational numbers m/n such that

$$0>\frac{m}{n}-\frac{L_1}{L_2}>-\frac{1}{n^2}.$$

This implies $nL_1/L_2-1<m<nL_1/L_2$ and thus $m=[nL_1/L_2]$. Proceeding exactly as in the proof of Theorem 1, the result follows. □

5. Case $\theta_1=\theta_2$ and $\sigma_1=-\sigma_3$. Proof of Theorem 3

We focus on the case $\theta_1=\theta_2=\theta_{k_1,k_2}^+$ and $\sigma_1=-\sigma_3=1$, which gives Theorem 3, leaving the analogous cases $\theta_1=\theta_2=\theta_{k_1,k_2}^-$ or $\sigma_1=-\sigma_3=-1$ to the interested reader. In such a case, the system in Equation (14) becomes

$$\begin{cases} (k_1,k_2)\in A \\ \frac{k_1}{\sqrt{2k_1^2-1}}\operatorname{cn}\left(\frac{\eta a_1}{\sqrt{2k_1^2-1}};k_1\right)=\frac{k_2}{\sqrt{2k_2^2-1}}\operatorname{cn}\left(\frac{\eta(L+a_2)}{\sqrt{2k_2^2-1}};k_2\right)=\operatorname{sech}(\eta a_3)=\theta_{k_1,k_2}^+ \\ \frac{k_1}{\sqrt{2k_1^2-1}}\operatorname{cn}\left(\frac{\eta(L_1+a_1)}{\sqrt{2k_1^2-1}};k_1\right)=\frac{k_2}{\sqrt{2k_2^2-1}}\operatorname{cn}\left(\frac{\eta(L_1+a_2)}{\sqrt{2k_2^2-1}};k_2\right)=\operatorname{sech}(\eta a_4)=\theta_{k_1,k_2}^+ \\ \sigma_2=-\sigma_4=1 \\ \begin{cases} k_1<k_2 \\ \sigma_5=\sigma_6=-1 \end{cases}\vee\begin{cases} k_1>k_2 \\ \sigma_5=\sigma_6=1 \end{cases}\vee\begin{cases} k_1=k_2 \\ a_3=a_4=0 \end{cases} \end{cases}$$

that is

$$\begin{cases} (k_1,k_2)\in A \\ \frac{k_1}{\sqrt{2k_1^2-1}}\operatorname{cn}\left(\frac{\eta a_1}{\sqrt{2k_1^2-1}};k_1\right)=\frac{k_2}{\sqrt{2k_2^2-1}}\operatorname{cn}\left(\frac{\eta(L+a_2)}{\sqrt{2k_2^2-1}};k_2\right)=\operatorname{sech}(\eta a_3)=\theta^+_{k_1,k_2} \\ \frac{k_1}{\sqrt{2k_1^2-1}}\operatorname{cn}\left(\frac{\eta(L_1+a_1)}{\sqrt{2k_1^2-1}};k_1\right)=\frac{k_2}{\sqrt{2k_2^2-1}}\operatorname{cn}\left(\frac{\eta(L_1+a_2)}{\sqrt{2k_2^2-1}};k_2\right)=\operatorname{sech}(\eta a_4)=\theta^+_{k_1,k_2} \\ \operatorname{sn}\left(\frac{\eta a_1}{\sqrt{2k_1^2-1}};k_1\right)>0,\quad \operatorname{sn}\left(\frac{\eta(L_1+a_1)}{\sqrt{2k_1^2-1}};k_1\right)<0 \\ \operatorname{sn}\left(\frac{\eta(L+a_2)}{\sqrt{2k_2^2-1}};k_2\right)>0,\quad \operatorname{sn}\left(\frac{\eta(L_1+a_2)}{\sqrt{2k_2^2-1}};k_2\right)<0 \\ \begin{cases} k_1<k_2 \\ \sigma_5=\sigma_6=-1 \end{cases} \vee \begin{cases} k_1>k_2 \\ \sigma_5=\sigma_6=1 \end{cases} \vee \begin{cases} k_1=k_2 \\ a_3=a_4=0 \end{cases} \end{cases} \tag{31}$$

Defining $\gamma_j\left(k_j,\eta,\theta\right)$ as in Section 3, we have that

$$\begin{cases} \frac{k_1}{\sqrt{2k_1^2-1}}\operatorname{cn}\left(\frac{\eta a_1}{\sqrt{2k_1^2-1}};k_1\right)=\theta^+_{k_1,k_2}, & \operatorname{sn}\left(\frac{\eta a_1}{\sqrt{2k_1^2-1}};k_1\right)>0 \\ \frac{k_1}{\sqrt{2k_1^2-1}}\operatorname{cn}\left(\frac{\eta(L_1+a_1)}{\sqrt{2k_1^2-1}};k_1\right)=\theta^+_{k_1,k_2}, & \operatorname{sn}\left(\frac{\eta(L_1+a_1)}{\sqrt{2k_1^2-1}};k_1\right)<0 \end{cases}$$

means

$$\begin{cases} a_1=\gamma_1\left(k_1,\eta,\theta^+_{k_1,k_2}\right) \\ L_1=mT_1(k_1,\eta)-2\gamma_1\left(k_1,\eta,\theta^+_{k_1,k_2}\right) \quad \text{for some } m\geq 1 \end{cases} \tag{32}$$

and

$$\begin{cases} \frac{k_2}{\sqrt{2k_2^2-1}}\operatorname{cn}\left(\frac{\eta(L+a_2)}{\sqrt{2k_2^2-1}};k_2\right)=\theta^+_{k_1,k_2}, & \operatorname{sn}\left(\frac{\eta(L+a_2)}{\sqrt{2k_2^2-1}};k_2\right)>0 \\ \frac{k_2}{\sqrt{2k_2^2-1}}\operatorname{cn}\left(\frac{\eta(L_1+a_2)}{\sqrt{2k_2^2-1}};k_2\right)=\theta^+_{k_1,k_2}, & \operatorname{sn}\left(\frac{\eta(L_1+a_2)}{\sqrt{2k_2^2-1}};k_2\right)<0 \end{cases}$$

means

$$\begin{cases} L_2=(n-m)\,T_2(k_2,\eta)+2\gamma_2\left(k_2,\eta,\theta^+_{k_1,k_2}\right) \quad \text{for some } n\geq m \\ a_2=\gamma_2\left(k_2,\eta,\theta^+_{k_1,k_2}\right)-L+pT_2(k_2,\eta) \quad \text{for some } p\geq n-m+1 \end{cases}$$

where m is the same integer of the system in Equation (32). Hence, the system in Equation (31) amounts to

$$\begin{cases} (k_1,k_2)\in A \\ L_1=mT_1(k_1,\eta)-2\gamma_1\left(k_1,\eta,\theta^+_{k_1,k_2}\right) \quad \text{for some } m\geq 1 \\ L_2=(n-m)\,T_2(k_2,\eta)+2\gamma_2\left(k_2,\eta,\theta^+_{k_1,k_2}\right) \quad \text{for some } n\geq m \\ a_1=\gamma_1\left(k_1,\eta,\theta^+_{k_1,k_2}\right) \\ a_2=\gamma_2\left(k_2,\eta,\theta^+_{k_1,k_2}\right)-L+pT_2(k_2,\eta) \quad \text{for some } p\geq n-m+1 \\ \operatorname{sech}(\eta a_3)=\operatorname{sech}(\eta a_4)=\theta^+_{k_1,k_2} \\ \begin{cases} k_1<k_2 \\ a_3,a_4<0 \end{cases} \vee \begin{cases} k_1>k_2 \\ a_3,a_4>0 \end{cases} \vee \begin{cases} k_1=k_2 \\ a_3=a_4=0. \end{cases} \end{cases} \tag{33}$$

Remark 5. *Suppose $L_1/L_2\notin\mathbb{Q}$. If we assume $k_1=k_2$ in the system in Equation (14), then we have $\theta_1=\theta_2=1$ and $\sigma_1=-\sigma_3$ (see Remark 4). Hence, a solution to the problem in. Equation (6) with plus* *sign gives rise to a solution to the system in Equation (33). On the other hand, a solution to the system in Equation (33) with* $k_1 = k_2$ *is such that* $L = L_1 + L_2 = nT$ *and* $a_2 = a_1 - L + pT = a_1 + (p-n)T$*, where* $T = T_1(k_1,\eta) = T_2(k_2,\eta)$*,* $a_1 \in (0, T/4)$ *and* $a_2 \in [0, T)$*. This forces* $p = n$ *and thus* $a_1 = a_2$*, so that the corresponding solution to the problem in Equation (6) is periodic on the circle.*

Now, recall that $T_j(k_j,\eta) := \dfrac{S(k_j)}{\eta}$. By the definition of $\gamma_j = \gamma_j\left(k_j,\eta,\theta^+_{k_1,k_2}\right)$, one has

$$\operatorname{cn}\left(\frac{\eta}{\sqrt{2k_j^2-1}}\gamma_j;k_j\right) = \frac{\sqrt{2k_j^2-1}}{k_j}\theta^+_{k_1,k_2} \tag{34}$$

with $\gamma_j \in (0, T_j/4)$. This implies

$$0 < \frac{\eta}{\sqrt{2k_j^2-1}}\gamma_j < \frac{\eta}{\sqrt{2k_j^2-1}}\frac{S(k_j)}{4\eta} = \frac{S(k_j)}{4\sqrt{2k_j^2-1}} = K(k_j)$$

and therefore Equation (34) yields that

$$\gamma_j\left(k_j,\eta,\theta^+_{k_1,k_2}\right) = \frac{\sqrt{2k_j^2-1}}{\eta}\operatorname{arccn}\left(\frac{\sqrt{2k_j^2-1}}{k_j}\theta^+_{k_1,k_2};k_j\right).$$

Hence, defining

$$\gamma(k_1,k_2) := \sqrt{2k_1^2-1}\operatorname{arccn}\left(\frac{\sqrt{2k_1^2-1}}{k_1}\theta^+_{k_1,k_2};k_1\right) = \sqrt{2k_1^2-1}\int_{\frac{\sqrt{2k_1^2-1}}{k_1}\theta^+_{k_1,k_2}}^{1}\frac{dt}{\sqrt{(1-t^2)\left(1-k_1^2(1-t^2)\right)}}$$

and observing that $\theta^+_{k_1,k_2} = \theta^+_{k_2,k_1}$, one has

$$\gamma_1\left(k_1,\eta,\theta^+_{k_1,k_2}\right) = \frac{1}{\eta}\gamma(k_1,k_2) \quad \text{and} \quad \gamma_2\left(k_2,\eta,\theta^+_{k_1,k_2}\right) = \frac{1}{\eta}\gamma(k_2,k_1).$$

Thus, the first three equations of the system in Equation (33) are equivalent to

$$\begin{cases} (k_1,k_2) \in A \\ \eta L_1 = mS(k_1) - 2\gamma(k_1,k_2) & \text{for some } m \geq 1 \\ \eta L_2 = (n-m)S(k_2) + 2\gamma(k_2,k_1) & \text{for some } n \geq m. \end{cases} \tag{35}$$

To prove Theorem 3, we use the following lemma, concerning the existence of a globally defined implicit function. Its proof is classical, so we leave it to the interested reader.

Lemma 3. *Let* $b_i \in \mathbb{R}$ *for* $i = 1,...,4$ *and let* $G : (b_1,b_2) \times (b_3,b_4) \to \mathbb{R}$ *be a continuous function such that for all* $x \in (b_1,b_2)$ *the following properties hold:*

- *the mapping* $G(x,\cdot)$ *is strictly increasing on* (b_3,b_4)*;*
- $\lim_{y\to b_3^+} G(x,y) < 0$ *and* $\lim_{y\to b_4^-} G(x,y) > 0$.

Then, the set of solutions to the equation $G(x,y) = 0$ *is the graph of a continuous function* $g\colon (b_1,b_2) \to (b_3,b_4)$*.*

Proof of Theorem 3. Let $n > m \geq 1$ and for $(k_1, k_2) \in A$ define the continuous functions

$$F_m(k_1, k_2) := mS(k_1) - 2\gamma(k_1, k_2) \quad \text{and} \quad F_{m,n}(k_1, k_2) := (n-m)\, S(k_2) + 2\gamma(k_2, k_1).$$

We also define F_m and $F_{m,n}$ on the segments $\left\{(k_1, 1) : \sqrt{3}/2 \leq k_1 < 1\right\}$ and $\left\{(1, k_2) : \sqrt{3}/2 \leq k_2 < 1\right\}$ of the boundary of A, respectively, where the above definitions also make sense.

Fix $\sqrt{3}/2 < \lambda < 1$ such that the square $Q = [\lambda, 1] \times [\lambda, 1]$ is contained into the closure of A and the partial derivatives $\partial F_1/\partial k_1$ and $\partial F_{1,2}/\partial k_2$ are strictly positive on Q. The existence of such a square can be checked by using the explicit expressions

$$F_1(k_1, k_2) = 2\sqrt{2k_1^2 - 1}\left(2K(k_1) - \int_{\frac{\sqrt{2k_1^2-1}}{k_1}\theta^+_{k_1,k_2}}^{1} \frac{dt}{\sqrt{(1-t^2)\left(1-k_1^2(1-t^2)\right)}}\right), \tag{36}$$

$$F_{1,2}(k_1, k_2) = 2\sqrt{2k_2^2 - 1}\left(2K(k_2) + \int_{\frac{\sqrt{2k_2^2-1}}{k_2}\theta^+_{k_1,k_2}}^{1} \frac{dt}{\sqrt{(1-t^2)\left(1-k_2^2(1-t^2)\right)}}\right), \tag{37}$$

where $\theta^+_{k_1,k_2}$ is given by Equation (18). Similarly, one checks that also F_1 is strictly positive on Q, while $F_{1,2}$ obviously is. Consequently, $\partial F_m/\partial k_1$, $\partial F_{m,n}/\partial k_2$, F_m and $F_{m,n}$ are also strictly positive on Q (recall that the function S is strictly increasing and positive). Define

$$\mu_m := \max_{\lambda \leq k_2 \leq 1} F_m(\lambda, k_2), \quad \mu_{m,n} := \max_{\lambda \leq k_1 \leq 1} F_{m,n}(k_1, \lambda) \quad \text{and} \quad \eta_{m,n} := \frac{\max\{\mu_m, \mu_{m,n}\}}{L_1},$$

and let $\eta > \eta_{m,n}$, so that $\eta L_2 > \eta L_1 > \max\{\mu_m, \mu_{m,n}\}$. By continuity of F_m and $F_{m,n}$, and using again the explicit expressions in Equations (36)–(37) (with general m and n inserted) as $k_1, k_2 \to 1$, we have that

$$\lim_{k_1 \to \lambda^+} F_m(k_1, k_2) = F_m(\lambda, k_2) \leq \mu_m < \eta L_1 \quad \text{and} \quad \lim_{k_1 \to 1^-} F_m(k_1, k_2) = +\infty$$

for every fixed $k_2 \in [\lambda, 1]$, and

$$\lim_{k_2 \to \lambda^+} F_{m,n}(k_1, k_2) = F_{m,n}(k_1, \lambda) \leq \mu_{m,n} < \eta L_2 \quad \text{and} \quad \lim_{k_2 \to 1^-} F_{m,n}(k_1, k_2) = +\infty$$

for every fixed $k_1 \in [\lambda, 1]$. Then, Lemma 3 ensures that the level sets

$$\{(k_1, k_2) \in Q : F_m(k_1, k_2) = \eta L_1\} \quad \text{and} \quad \{(k_1, k_2) \in Q : F_{m,n}(k_1, k_2) = \eta L_2\}$$

respectively, are the graphs $k_1 = f(k_2)$ and $k_2 = g(k_1)$ of two continuous functions f, g defined on $[\lambda, 1]$. The first graph joins a point on the segment $[\lambda, 1] \times \{1\}$ to a point on $[\lambda, 1] \times \{\lambda\}$, the latter one joins a point on $\{\lambda\} \times [\lambda, 1]$ to a point on $\{1\} \times [\lambda, 1]$, and therefore the two level sets must intersect in the interior of Q at a point (k_1, k_2), which thus solves the system in Equation (35). Then, Lines 4–7 of the system in Equation (33) fix the values of a_1, a_2, a_3, a_4, by taking p as the unique integer such that the corresponding a_4 belongs to $(0, T_2]$. This completes the proof. □

Remark 6. *In the proof of Theorem 3, the sign of the function F_1 can be easily checked. Indeed, taking into account that $\theta^+_{k_1,k_2} \geq 1/\sqrt{2}$, one has*

$$F_1(k_1, k_2) > 2\sqrt{2k_1^2 - 1}\int_{\frac{\sqrt{2k_1^2-1}}{k_1\sqrt{2}}}^{1} \frac{1}{\sqrt{1-t^2}}\left(\frac{1}{\sqrt{1-k_1^2 t^2}} - \frac{1}{\sqrt{1-k_1^2(1-t^2)}}\right) > 0.$$

On the contrary, the analysis of the sign of $\partial F_1/\partial k_1$ and $\partial F_{1,2}/\partial k_2$ over the set A is rather involved and we could not perform it exactly. Therefore, we based our argument concerning the existence of the square Q on the numerical evidence given by the plots of their graphs (see Figure 5), for which we used the software `Wolfram MATHEMATICA 10.4.1`.

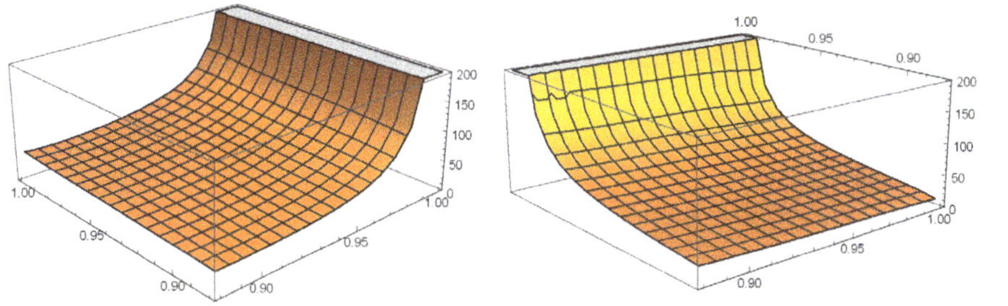

Figure 5. The functions $\partial F_1/\partial k_1$ and $\partial F_{1,2}/\partial k_2$ over the square $[\lambda, 1]^2$ with $\lambda = 0.88$.

Author Contributions: All the authors contribute equally to this work

Funding: The research of D.N. and S.R. was funded in part by Departmental Project 2018-CONT-0127, University of Milano Bicocca.

Conflicts of Interest: The authors declare no conflict of interest. The funders had no role in the design of the study; in the collection, analyses, or interpretation of data and in the writing of the manuscript, or in the decision to publish the results.

References

1. Adami, R.; Serra, E.; Tilli, P. Nonlinear dynamics on branched structures and networks. *Riv. Mat. Univ. Parma* **2017**, *8*, 109–159.
2. Cacciapuoti, C.; Finco, D.; Noja, D. Ground state and orbital stability for the NLS equation on a general starlike graph with potentials. *Nonlinearity* **2017**, *30*, 3271–3303. [CrossRef]
3. Noja, D. Nonlinear Schrödinger equation on graphs: recent results and open problems. *Philos. Trans. R. Soc. A* **2014**, *372*, 20130002. [CrossRef] [PubMed]
4. Gnutzmann, S.; Waltner, D. Stationary waves on nonlinear quantum graphs. I. General framework and canonical perturbation theory, *Phys. Rev. E* **2016**, *93*, 032204. [PubMed]
5. Gnutzmann, S.; Waltner, D. Stationary waves on nonlinear quantum graphs. II. Application of canonical perturbation theory in basic graph structures. *Phys. Rev. E* **2016**, *94*, 062216. [CrossRef] [PubMed]
6. Marzuola, J.; Pelinovsky, D.E. Ground states on the dumbbell graph. *Appl. Math. Res. Express* **2016** *1*, 98–145. [CrossRef]
7. Pelinovsky, D.E.; Schneider, G. Bifurcations of standing localized waves on periodic graphs. *Ann. Henri Poincaré* **2017**, *18*, 1185. [CrossRef]
8. Sobirov, Z.; Matrasulov, D.U.; Sabirov, K.K.; Sawada, S.; Nakamura, K. Integrable nonlinear Schrödinger equation on simple networks: Connection formula at vertices. *Phys. Rev. E* **2010**, *81*, 066602. [CrossRef] [PubMed]
9. Sabirov, K.K.; Sobirov, Z.A.; Babajanov, D.; Matrasulov, D.U. Stationary nonlinear Schrödinger equation on simplest graphs. *Phys. Lett. A* **2013**, *377*, 860–865. [CrossRef]
10. Adami, R.; Cacciapuoti, C.; Finco, D.; Noja, D. Stable standing waves for a NLS on star graphs as local minimizers of the constrained energy. *J. Differ. Equ.* **2016**, *260*, 7397–7415,
11. Adami, R.; Serra, E.; Tilli, P. NLS ground states on Graphs. *Calc. Var. Partial Differ. Equ.* **2015**, *54*, 743–761. [CrossRef]
12. Adami, R.; Serra, E.; Tilli, P. Threshold phenomena and existence results for NLS ground states on metric graphs. *J. Funct. Anal.* **2016**, *271*, 201–223. [CrossRef]

13. Adami, R.; Serra, E.; Tilli, P. Negative Energy Ground States for the L2-Critical NLSE on Metric Graphs. *Comm. Math. Phys.* **2017**, *352*, 387–406. [CrossRef]
14. Berkolaiko, G.; Kuchment, P. *Introduction to Quantum Graphs;* Mathematical Surveys and Monographs 186; American Mathematical Society: Providence, RI, USA, 2013.
15. Noja, D.; Rolando, S.; Secchi, S. Standing waves for the NLS on the double–bridge graph and a rational-irrational dichotomy. *J. Differ. Equ.* **2019**, *451*, 147–178. [CrossRef]
16. Adami, R.; Serra, E.; Tilli, P. Multiple positive bound states for the subcritical NLS equation on metric graphs. *Calc. Var. Partial Differ. Equ.* **2019**, *58*. [CrossRef]
17. Cacciapuoti, C.; Finco, D.; Noja, D. Topology-induced bifurcations for the nonlinear Schrödinger equation on the tadpole graph. *Phys. Rev. E* **2015**, *91*, 013206. [CrossRef] [PubMed]
18. Noja, D.; Pelinovsky, D.; Shaikhova, G. Bifurcation and stability of standing waves in the nonlinear Schrödinger equation on the tadpole graph. *Nonlinearity* **2015**, *28*, 2343–2378. [CrossRef]
19. Lawden, D.F. *Elliptic Functions and Applications;* Springer: New York, NY, USA, 1989.
20. Olver, F.W.J.; Lozier, D.W.; Boisvert, R.F.; Clark, C.W. *NIST Handbook of Mathematical Functions;* Cambridge University Press: New York, NY, USA, 2010.
21. Niven, I. *Diophantine Approximations;* Interscience Tracts in Pure and Applied Mathematics No. 14; Interscience Publishers, Wiley & Sons: New York, NY, USA, 1963 .

symmetry

Article

On the Nodal Structure of Nonlinear Stationary Waves on Star Graphs

Ram Band [1], Sven Gnutzmann [2,*] and August J. Krueger [3]

1 Department of Mathematics, Technion—Israel Institute of Technology, Haifa 3200003, Israel; ramband@technion.ac.il
2 School of Mathematical Sciences, University of Nottingham, Nottingham NG7 2RD, UK
3 Department of Mathematics, Rutgers University, Piscataway, NJ 08854-8019, USA; akrueger@math.rutgers.edu
* Correspondence: sven.gnutzmann@nottingham.ac.uk

Received: 14 January 2019; Accepted: 01 February 2019; Published: 5 February 2019

Abstract: We consider stationary waves on nonlinear quantum star graphs, i.e., solutions to the stationary (cubic) nonlinear Schrödinger equation on a metric star graph with Kirchhoff matching conditions at the centre. We prove the existence of solutions that vanish at the centre of the star and classify them according to the nodal structure on each edge (i.e., the number of nodal domains or nodal points that the solution has on each edge). We discuss the relevance of these solutions in more applied settings as starting points for numerical calculations of spectral curves and put our results into the wider context of nodal counting, such as the classic Sturm oscillation theorem.

Keywords: quantum graphs; nonlinear Schrödinger equation; nodal structure

1. Introduction

Sturm's oscillation theorem [1] is a classic example for how solutions of linear self-adjoint differential eigenvalue problems $D\phi(x) = \lambda\phi(x)$ (where D is a Sturm-Liouville operator) are ordered and classified by the number of nodal points. According to Sturm's oscillation theorem, the n-th eigenfunction, ϕ_n, has $n-1$ nodal points, when the eigenfunctions are ordered by increasing order of their corresponding eigenvalues $\lambda_1 < \lambda_2 < \lambda_3 < \ldots$. Equivalently, the number ν_n of nodal domains (the connected domains where ϕ_n has the same sign) obeys $\nu_n = n$ for all n.

In higher dimensions, (e.g., for the free Schrödinger equation $-\Delta\phi(\mathbf{x}) = \lambda\phi(\mathbf{x})$ on a bounded domain with self-adjoint boundary conditions) the number of nodal domains is bounded from above, $\nu_n \leq n$, by Courant's theorem [2] (see [3] for the case of Schrödinger equation with potential). Furthermore, there is only a finite number of Courant sharp eigenfunctions for which $\nu_n = n$, as was shown by Pleijel [4].

In (linear) quantum graph theory one considers the Schrödinger equation with self-adjoint matching conditions at the vertices of a metric graph. Locally, graphs are one-dimensional though the connectivity of the graph allows to mimic some features of higher dimensions. Nodal counts for quantum graphs have been considered for more than a decade [5]. For example, it has been shown in [5] that Courant's bound applies to quantum graphs as well. Yet, for graphs there are generically infinitely many Courant sharp eigenfunctions [6,7]. For tree graphs it has been proven that all generic eigenfunctions are Courant sharp, i.e., $\nu_n = n$ [8,9] . In other words, Sturm's oscillation theorem generalizes to metric trees graphs. It has been proven by one of us that the converse also holds, namely that if the graph's nodal count obeys $\nu_n = n$ for all n, then this graph is a tree [10]. When a graph is not a tree, its first Betti number, $\beta := E - V + 1$ is positive. Here, E, V are correspondingly the numbers of graph's edges and vertices and β indicates the number of the graphs independent cycles. In addition to Courant's bound, the nodal count of a graph is bounded from below, $\nu_n \leq n - \beta$ as was shown first in [11]. The actual number of nodal domains may be characterized by various variational methods [12,13]. Some statistical properties of the nodal count are also known [6], but to date there is no general explicit formula or a full statistical description of the nodal count.

In the present work we present some related results concerning nodal points and nodal domains for *nonlinear* star graphs (see Figure 1).

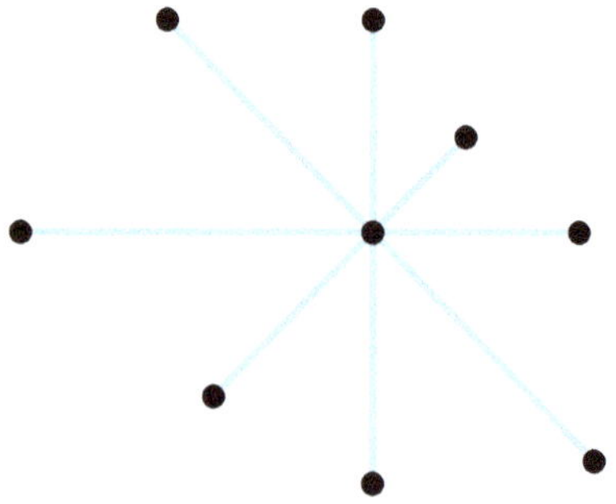

Figure 1. A star graph with $E = 8$ edges and $E + 1 = 9$ vertices.

Nonlinear wave equations on metric graphs (i.e., nonlinear quantum graphs) have recently attracted considerable interest both from the mathematical perspective and the applied regime. They allow the study of intricate interplay between the non-trivial connectivity and the nonlinearity. Among the physical applications of nonlinear wave equations on metric graphs is light transmission through a network of optical fibres or Bose-Einstein condensates in quasi one-dimensional traps. We refer to [14,15] where a detailed overview of the recent literature and some applications is given and just summarise here the relevant work related to the nodal counting. In a previous work some of us have shown that Sturm's oscillation theorem is generically broken for nonlinear quantum stars, apart from the special case of an interval [16]. This is not unexpected as the set of solutions is known to have a far more complex structure. Our main result here is that the nonlinear case of a metric star allows for solutions with any given number of nodal domains on each edge. Namely, for a star with E edges and a certain E-tuple, $(n_1, \ldots, n_E)$ of non-negative integers there are solutions with n_e nodal points on the e-th edge for $e = 1, \ldots, E$.

In the remainder of the introduction chapter we define the setting. In Section 2 we state our main results. In Section 3 we present the nonlinear generalization of Sturm's oscillation theorem to an interval, some general background and properties of nonlinear solutions as well as a few motivating numerical results. In Section 4 we prove the main theorems and afterwards in Section 5 we discuss our results and their possible implications in the broader context of nonlinear quantum graphs.

1.1. The Setting—Nonlinear Star Graphs

Metric star graphs are a special class of metric trees with E edges and $E + 1$ vertices such that all edges are incident to one common vertex (see Figure 1). The common vertex will be called the centre of the star and the other vertices will be called the boundary. We assume that each edge has a finite length $0 < \ell_e < \infty$ ($e = 1, \ldots, E$) and a coordinate $x_e \in [0, \ell_e]$ such that $x_e = 0$ at the centre and $x_e = \ell_e$ at the boundary. On each edge $e = 1, \ldots, E$ we consider the stationary nonlinear Schrödinger (NLS) equation

$$-\frac{d^2}{dx_e^2}\phi_e(x_e) + g|\phi_e(x_e)|^2\phi_e(x_e) = \mu\,\phi_e(x_e) \tag{1}$$

for $\phi_e : [0, \ell_e] \to \mathbb{C}$. Here g is a nonlinear coupling parameter and μ a spectral parameter. We consider this as a generalized eigenequation with eigenvalues μ. We have assumed here that the nonlinear interaction is homogeneously repulsive ($g > 0$) or attractive ($g < 0$) on all edges and will continue to do so throughout this manuscript. One may consider more general graphs where g takes different values (and different signs) on different edges (or even where $g \to g_e(x_e)$ is a real scalar function on the graph). It is not, however, our aim to be as general as possible. In the following we restrict ourselves to one generic setting in order to keep the notation and discussion as clear and short as possible. We will later discuss some straightforward generalizations of our results.

At the centre we prescribe Kirchhoff (a.k.a Neumann) matching conditions

$$\phi_e(0) = \phi_{e'}(0) \qquad \text{for all } 1 \leq e < e' \leq E \tag{2}$$

$$\sum_{e=1}^{E} \frac{d\phi_e}{dx_e}(0) = 0 \tag{3}$$

and at the boundary we prescribe Dirichlet conditions $\phi_e(\ell_e) = 0$.

For the coupling constant g it is sufficient without loss of generality to consider three different cases. For $g = 0$ one recovers the linear Schrödinger equation and thus standard quantum star graphs. For $g = 1$ one has a nonlinear quantum star graph with repulsive interaction and for $g = -1$ one has a nonlinear quantum star with attractive interaction. If g takes any other non-zero value a simple rescaling $\phi_e(x_e) \mapsto \frac{1}{\sqrt{|g|}}\phi_e(x_e)$ of the wavefunction is equivalent to replacing $g \mapsto \frac{g}{|g|} = \pm 1$.

Without loss of generality we may focus on real-valued and twice differentiable solutions $\{\phi_e(x_e)\}_{e=1}^{E}$, where twice differentiable refers separately to each $\phi_e : (0, \ell_e) \to \mathbb{C}$. We also assume that the solution is not the constant zero function on the graph, namely that there is an edge e and some point $\hat{x}_e \in [0, \ell_e]$ with $\phi_e(\hat{x}_e) \neq 0$. Moreover, any complex-valued solution is related to a real-valued solution by a global gauge-transformation (i.e., a change of phase $\phi_e(x_e) \mapsto \phi_e(x_e)e^{i\alpha}$) [14].

1.2. The Nodal Structure

We will call a solution $\{\phi_e(x_e)\}_{e=1}^{N}$ *regular* if the wavefunction does not vanish on any edge, that is for each edge e there is $\hat{x}_e \in (0, \ell_e)$ with $\phi_e(\hat{x}_e) \neq 0$. Accordingly, *non-regular* solutions vanish identically on some edges, in other words there is (at least) one edge e such that $\phi_e(x_e) = 0$ for all $x_e \in [0, \ell_e]$.

A solution with a node at the centre, $\phi_e(0) = 0$ (by continuity this is either true for all e or for none) will be called *central Dirichlet* because it satisfies Dirichlet conditions at the centre (in addition to the Kirchhoff condition). Hence, non-regular solutions are always central Dirichlet. Our main theorem will construct solutions which are regular and central Dirichlet. Note that from a regular central Dirichlet solution on a metric star graph G one can construct non-regular solutions on a larger metric star graph G', if G is a metric subgraph of G': on each edge $e \in G' \setminus G$ one may just extend the solution by setting $\phi_e(x_e) = 0$ for all $x_e \in [0, \ell_e]$.

Our main aim is to characterize solutions in terms of their nodal structure. The nodal structure is described in terms of either the number ν of nodal domains (maximal connected subgraphs where $\phi_e(x_e) \neq 0$) or by the number ξ of nodal points. We will include in the count the trivial nodal points at the boundary. Note that regular solutions which are not central Dirichlet obey $\nu = \xi + 1 - E$ while regular central Dirichlet solutions obey $\nu = \xi - 1$. We have stated in the introduction that in the linear case, $g = 0$, such a characterization is very well understood even for the more general tree graphs, which obey a generalized version of Sturm's oscillation theorem.

As we will see, the solutions of nonlinear star graphs have a very rich structure and a classification of solutions in terms of the total numbers ν or ξ of nodal domains or nodal points is far from being unique. We will thus use a more detailed description of the nodal structure of the solutions. To each regular solution $\{\phi_e(x_e)\}_{e=1}^{E}$ we associate the E-tuple

$$\mathbf{n} = (n_1, \ldots, n_E) \in \mathbb{N}^E \tag{4}$$

where $n_e \geq 1$ is the number of nodal domains of the wavefunction $\phi_e(x_e)$ on the edge $x_e \in [0, \ell_E]$. For solutions which are not central Dirichlet, n_e also equals the number of nodal points of $\phi_e(x_e)$

(including the nodal point at the boundary). We will call $\mathbf{n} \in \mathbb{N}^E$ the *(regular) nodal edge count structure* of the (regular) solution $\{\phi_e(x_e)\}_{e=1}^N$. For non-regular solutions one may characterize the nodal structure in a similar way by formally setting $n_e = \infty$ for all edges where the wavefunction is identical zero. In that case we speak of a *non-regular nodal edge count structure*. Note that we do not claim that the nodal edge count structure, $\mathbf{n}$, leads to a unique characterization of the solutions (which actually come in one-parameter families). Indeed we have numerical counter-examples. With this more detailed description we show that a much larger set of nodal structures is possible in nonlinear quantum star graphs compared to the linear case, as is stated in the next section.

2. Statement of Main Theorems

Our main results concern the existence of solutions with any given nodal edge structure. We state two theorems: One for repulsive nonlinear interaction $g = 1$ and one for attractive nonlinear interaction $g = -1$. The two theorems establish the existence of central Dirichlet solutions with nodal edge structure $\mathbf{n} = (1, \ldots, 1)$ subject to (achievable) conditions on the edge lengths. As corollaries, we get the existence of central Dirichlet solutions with any prescribed values of $\mathbf{n}$ (again subject to some achievable conditions on the lengths). Throughout this section we consider a nonlinear quantum star graph as described in Section 1.1. In order to avoid trivial special cases we will assume $E \geq 3$. Indeed, $E = 1$ is the interval and well understood and $E = 2$ reduces to an interval (of total length $\ell_1 + \ell_2$) as the Kirchhoff vertex condition in this case just states that the wavefunction is continuous and has a continuous first derivative. We will also assume that all edge lengths are different. Without loss of generality we take them as ordered $\ell_e < \ell_{e+1}$ $(e = 1, \ldots, E-1)$.

Theorem 1. *If $g = 1$ (repulsive case) and either*

1. *the number of edges E is odd, or*
2. *E is even and*

$$\sqrt{\frac{m_+}{m_-}\frac{1+m_-}{1+m_+}} > \frac{E}{E-2}, \tag{5}$$

where $0 < m_- < m_+ < 1$ are implicitly defined in terms of the edge lengths l_1, $l_{\frac{E}{2}+1}$, $l_{\frac{E}{2}+2}$ by

$$\begin{aligned} K(m_+)\sqrt{1+m_+} &= \frac{\pi}{2}\frac{\ell_{\frac{E}{2}+2}}{\ell_1}, \\ K(m_-)\sqrt{1+m_-} &= \frac{\pi}{2}\frac{\ell_{\frac{E}{2}+1}}{\ell_1}, \end{aligned} \tag{6}$$

with

$$K(m) = \int_0^1 \frac{1}{\sqrt{1-u^2}\sqrt{1-mu^2}}du \tag{7}$$

being the complete elliptic integral of first kind,

then there exists a regular central Dirichlet solution for some positive value of the spectral parameter $\mu = k^2 > \frac{\pi^2}{\ell_1^2}$ such that there is exactly one nodal domain on each edge, i.e., the nodal edge structure $\mathbf{n}$ satisfies $n_e = 1$ for all edges e.

Note that the condition in this theorem for even number of edges involves only three edge lengths and can be stated in terms of two ratios that satisfy $\frac{\ell_{\frac{E}{2}+2}}{\ell_1} \geq \frac{\ell_{\frac{E}{2}+1}}{\ell_1} \geq 1$ (as we have ordered the edges by lengths). If the larger ratio $\frac{\ell_{\frac{E}{2}+2}}{\ell_1}$ is given then one may always achieve this condition by choosing the other ratio sufficiently small (as $\frac{\ell_{\frac{E}{2}+1}}{\ell_1} \to 1$ one has $m_- \to 0$ and the left-hand side of condition (5) grows without any bound). Figure 2 shows a graph of the regions where the two length ratios satisfy condition (5) for star graphs with E edges. One can see how the condition becomes less restrictive when the number of edges is large. We will present the proof of Theorem 1 in Section 4.1. The proof shows that the condition (5) is not optimal. Less restrictive conditions that depend on other edge lengths may be stated. Nevertheless, we have chosen to state the condition (5), as its form is probably more compactly phrased than other conditions would be.

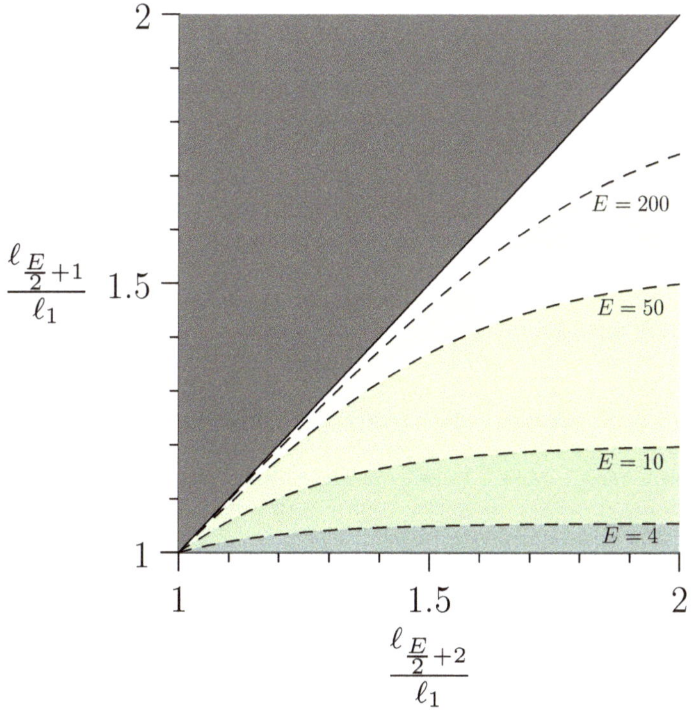

Figure 2. The shaded regions indicate choices of relative edge lengths $1 < \frac{\ell_{\frac{E}{2}+1}}{\ell_1} < \frac{\ell_{\frac{E}{2}+2}}{\ell_1}$ that satisfy the condition (5) of Theorem 1. The dashed lines indicate the boundary of regions for a star graph with E edges (where $E = 4, 10, 50, 200$). Condition (5) is satisfied below the dashed lines.

Theorem 2. *Let $g = -1$ (attractive case). If there exists an integer $M < E/2$ such that*

$$\sum_{e=M+1}^{E-1} \frac{1}{\ell_e^2} < \sum_{e=1}^{M} \frac{1}{\ell_e^2} < \sum_{e=M+1}^{E} \frac{1}{\ell_e^2}, \tag{8}$$

then there exists a regular central Dirichlet solution for some positive value of the spectral parameter $\mu = k^2 \in \left(0, \frac{\pi^2}{\ell_E^2}\right)$ such that there is exactly one nodal domain on each edge, i.e., the nodal edge structure $\mathbf{n}$ satisfies $n_e = 1$ for all edges e.

We will prove this theorem in Section 4.2. One may extend Theorem 2 to find negative values of the spectral parameter $\mu < 0$ under appropriate conditions on edge lengths using similar ideas as the ones used in our proof for $\mu > 0$. To keep the paper concise we focus here on $\mu > 0$.

In order to demonstrate how the two conditions (8) in Theorem 2 may be achieved, we point out that the following weaker conditions

$$\begin{aligned} \frac{\ell_M}{\ell_{M+1}} &< \sqrt{\frac{M}{E-M-1}} \\ \frac{\ell_1}{\ell_E} &> \sqrt{\frac{M}{E-M}} \end{aligned} \tag{9}$$

imply (8) (recalling that $\ell_1 < \ldots < \ell_E$). The conditions (9) are easy to apply and they may be achieved straight-forwardly. For instance, if E is odd and $M = (E-1)/2$ (the largest possible value for M) then the first inequality in (9) is always satisfied and the second condition gives the restriction $1 > \frac{\ell_1}{\ell_E} > \sqrt{\frac{M}{M+1}}$ on the ratio between the smallest and largest edge length. In addition to that one may easily construct a star graph with edge lengths which satisfy conditions (9) above. This is done by starting from a star graph which has only two different edge lengths $\ell_- < \ell_+$ where $\ell_e = \ell_-$ for $1 \le e \le M$ and $\ell_e = \ell_+$ for $M+1 \le e \le E$. If one chooses the ratio of the lengths in the range $\sqrt{\frac{M}{E-M}} < \frac{\ell_-}{\ell_+} < \sqrt{\frac{M}{E-M-1}}$ and then perturbs all edge lengths slightly to make them different then condition (9) is satisfied. Note however that, just as in Theorem 1, even the condition which is stated in Theorem 2 is not optimal and more detailed conditions can be derived from our proof in Section 4.2.

Before discussing some straight-forward implications let us also state here that the assumption that all edge lengths are different that we made for both Theorems 1 and 2 may be relaxed. This is because any two edges with the same length decouple in a certain way from the remaining graph. If one deletes pairs of edges of equal length from the graph until all edges in the remaining graph are different one may apply the theorems to the remaining graph (if the remaining graph has at least three edges). This will be discussed more in Remark 1.

In the remainder of this section we discuss the implications of the two theorems for finding solutions with a given nodal edge structure $\mathbf{n} \in \mathbb{N}^E$. In this case we divide each edge length into n_e fractions $\ell_e = n_e \tilde{\ell}_e$. The n-th fraction $\tilde{\ell}_e$ then corresponds to the length of one nodal domain. For the rest of this section we do not assume that the edge lengths $\{\ell_e\}$ are ordered by length and different, rather we now assume that these assumptions apply to the fractions, i.e., $\tilde{\ell}_e < \tilde{\ell_{e+1}}$ $(e = 1, \ldots, E-1)$. By first considering the metrically smaller star graph with edge lengths $\{\tilde{\ell}_e\}$ Theorems 1 and 2 establish the existence of solutions on this smaller graph subject to conditions on the lengths $\{\tilde{\ell}_e\}$. These solutions can be extended straight-forwardly to a solution on the full star graph. Indeed, as we explain in more detail in Section 3.1, the solution on each edge is a naturally periodic function given by an elliptic deformation of a sine and shares the same symmetry around nodes and extrema, as the sine function. The main relevant difference to a sine is that the period of the solution depends on the amplitude. In the repulsive case one then obtains the following.

Corollary 1. *Let $g = 1$ (repulsive case) and $\mathbf{n} \in N^E$. If either*

1. *E is odd, or*
2. *E is even and the fractions $\tilde{\ell}_e = \ell_e / n_e$ $(e = 1, \ldots, E)$ satisfy the condition (5),*

then there exists a regular central Dirichlet solution for some positive value of the spectral parameter $\mu = k^2 > \frac{\pi^2}{\tilde{\ell}_1^2}$ 'with regular nodal edge count structure $\mathbf{n}$.

Similarly, Theorem 2 implies the following.

Corollary 2. *Let $g = -1$ (attractive case) and $\mathbf{n} \in \mathbb{N}^E$. If the fractions $\tilde{\ell}_e = \ell_e / n_e$ satisfy condition (8) then there exists a regular central Dirichlet solution for some positive value of the spectral parameter $\mu = k^2 \in \left(0, \frac{\pi^2}{\tilde{\ell}_E^2}\right)$ with regular nodal edge count structure $\mathbf{n}$.*

The corollaries above provide sufficient conditions for the existence of a central Dirichlet solution with a particular given nodal edge count. In addition to that, it is straight-forward to apply Theorems 1 and 2 to show that for any choice of edge lengths there are infinitely many E-tuples which can serve as the graph's regular central Dirichlet nodal structure.

Moreover Theorems 1 and 2 also imply infinitely many values for non-regular nodal structures, as every non-regular solution is equivalent to a regular solution on a subgraph.

Finally, we note that the proofs of Theorems 1 and 2 in Section 4 are constructive and they specify the corresponding solution up to a single parameter (which one may take to be $k = \sqrt{\mu}$) that may easily be found numerically.

3. General Background on the Solutions of Nonlinear Quantum Star Graphs

Before we turn to the proof of Theorems 1 and 2 we would like discuss how the implied regular central Dirichlet solutions are related to the complete set of solutions of the nonlinear star graph. Though we are far from having a full understanding of all solutions we can give a heuristic picture.

3.1. The Nonlinear Interval - Solutions and Spectral Curves

Let us start with giving a complete overview of the solutions for the interval (i.e., the star graph with $E = 1$). While these are well known and understood they play a central part in the construction of central Dirichlet solutions for star graphs in our later proof and serve as a good way to introduce some general background. On the half line $x \geq 0$ with a Dirichlet condition $\phi(0) = 0$ at the origin it is straight-forward to check (see also [14]) that the solutions for positive spectral parameters $\mu = k^2$ (where $k > 0$) are of the form

$$\phi(x) = \begin{cases} \chi^{(+)}_{m,k}(x) = k\sqrt{\frac{2m}{1+m}}\,\mathrm{sn}\left(\frac{kx}{\sqrt{1+m}}, m\right) & \text{in the repulsive case } g = 1\,, \\ \chi^{(-)}_{m,k}(x) = k\sqrt{\frac{2m(1-m)}{1-2m}}\,\frac{\mathrm{sn}\left(\frac{kx}{\sqrt{1-2m}}, m\right)}{\mathrm{dn}\left(\frac{kx}{\sqrt{1-2m}}, m\right)} & \text{in the attractive case } g = -1. \end{cases} \tag{10}$$

Here $\mathrm{sn}(y, m)$ and $\mathrm{dn}(y, m)$ are Jacobi elliptic functions with a deformation parameter m. The definition of elliptic functions allows m to take arbitrary values in the interval $m \in [0, 1]$ (as there are many conventions for these functions we summarize ours in Appendix A). Note that $\mathrm{sn}(y, m)$ is a deformed variant of the sine function and $\mathrm{sn}(y, 0) = \sin(y)$ and $\mathrm{dn}(y, 0) = 1$.

For any spectral parameter $\mu = k^2$ there is a one-parameter family of solutions parameterised by the deformation parameter m. In the repulsive case the deformation parameter may take values $m \in (0, 1]$ (as for $m = 0$ one obtains the trivial solution $\chi^{(+)}_{0,k}(x) = 0$) and in the attractive case $m \in \left(0, \frac{1}{2}\right)$ (the expressions are not well defined for $m = \frac{1}{2}$ and for $m > \frac{1}{2}$ the expressions are no longer real).

Let us now summarise some properties of these solutions in the following proposition for the solutions of the NLS equation on the half line.

Proposition 1. *The solutions $\phi(x) = \chi^{(\pm)}_{m,k}(x)$ given in Equation (10) have the following properties*

1. *All solutions are periodic $\chi^{(\pm)}_{m,k}(x) = \chi^{(\pm)}_{m,k}(x + \Lambda^{(\pm)}(m,k))$ with a nonlinear wavelength*

$$\begin{aligned} \Lambda^{(+)}(m,k) &= \frac{4\sqrt{1+m}K(m)}{k} \\ \Lambda^{(-)}(m,k) &= \frac{4\sqrt{1-2m}K(m)}{k}, \end{aligned} \tag{11}$$

where $K(m)$ is the complete elliptic integral of first kind, Equation (7).

2. *For $m \to 0$ one regains the standard relation $\Lambda^{(\pm)}(0,k) = \frac{2\pi}{k}$ for the free linear Schrödinger equation. In the repulsive case $\Lambda^{(+)}(m,k)$ is an increasing function of m (at fixed k) that increases without bound as $m \to 1$. In the attractive case $\Lambda^{(-)}(m,k)$ is a decreasing function of m (at fixed k) with $\Lambda^{(-)}\left(\frac{1}{2},k\right) = 0$.*
3. *The nodal points are separated by half the nonlinear wavelength. Namely, $\chi^{(\pm)}_{m,k}(n\Lambda^{(\pm)}(m,k)/2) = 0$ for $n = 0, 1, , \ldots$.*
4. *The solutions are anti-symmetric around each nodal point and symmetric around each extremum, i.e., it has the same symmetry properties as a sine function.*
5. *As $\mathrm{sn}\,(K(m),m) = 1$ and $\mathrm{dn}\,(K(m),m) = \sqrt{1-m}$ the amplitude*

$$A^{(\pm)}(k,m) = \max\left(\chi^{(\pm)}_{m,k}(x)\right)_{x\geq 0} = \chi^{(\pm)}_{m,k}\left(\frac{\Lambda^{(\pm)}(m,k)}{4}\right)$$

   is given by

$$\begin{aligned} A^{(+)}(m,k) =& k\sqrt{\frac{2m}{1+m}} \\ A^{(-)}(m,k) =& k\sqrt{\frac{2m}{1-2m}}. \end{aligned} \tag{12}$$

6. *As $m \to 0^+$ the amplitude of the solutions also decreases to zero $A^{(\pm)}(0,k) = 0$ for both the repulsive and the attractive case. In this case the effective strength of the nonlinear interaction becomes weaker and the oscillations are closer. In the repulsive case the amplitude remains bounded as $m \to 1$ with $A^{(+)}(1,k) = k$. In the attractive case $A^{(-)}(m,k)$ grows without bound as $m \to \frac{1}{2}$.*

All statements in this proposition follow straight-forwardly from the known properties of elliptic integrals and elliptic functions and we thus omit the proof here. Furthermore, some of the statements in the proposition are mentioned explicitly in [14,16,17] and others follow easily from the definitions as given in the Appendix A.

For the NLS equation for $\phi(x)$ on an interval $x \in [0,\ell]$ with Dirichlet conditions at both boundaries $\phi(0) = \phi(\ell) = 0$ one obtains a full set of solutions straight-forwardly from the solutions $\chi^{(\pm)}_{m,k}(x)$ on the half-line by requiring that there is a nodal point at $x = \ell$. Since the distance between two nodal points in $\chi^{(\pm)}_{m,k}(x)$ is $\Lambda^{(\pm)}(k,m)/2$ the length of the interval has to be an integer multiple of half the nonlinear wavelength

$$2\ell = n\Lambda^{(\pm)}(k,m), \tag{13}$$

where the positive integer n is the number of nodal domains. We arrive at the following proposition.

Proposition 2. *The NLS Equation (1) on an interval of length ℓ with Dirichlet boundary conditions has a one-parameter family of real-valued solutions with n nodal domains, for each $n \in \mathbb{N}$. The relation between the spectral parameter $\mu = k^2$ and the deformation parameter m is dictated by Equation (13) and may be explicitly written as*

$$\begin{aligned} k^{(+)}_{n,\ell}(m) =& \frac{2n\sqrt{1+m}K(m)}{\ell} \\ k^{(-)}_{n,\ell}(m) =& \frac{2n\sqrt{1-2m}K(m)}{\ell} \end{aligned}. \tag{14}$$

We refer to $k^{(\pm)}_{n\ell}(m)$ (or its implicitly defined inverse $m^{(\pm)}_{n,\ell}(k)$) as spectral curves. As $k^{(\pm)}_{n+1,\ell}(m) > k^{(\pm)}_{n,\ell}(m)$, the spectral curves never cross (see Figure 3) and we obtain the first nonlinear generalization of Sturm's oscillation theorem as a corollary (see also Theorem 2.4 in [16]).

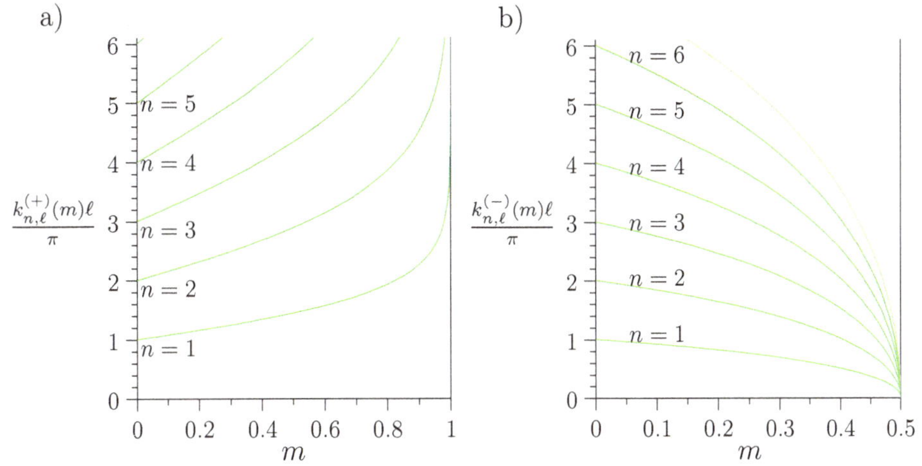

Figure 3. Spectral curves $k_{n,\ell}^{(\pm)}(m)$ for the repulsive (**a**) and attractive case (**b**). The n-th curve is obtained from the curve for $n=1$ by rescaling $k_{n,\ell}^{(\pm)}(m) = nk_{1,\ell}^{(\pm)}(m)$.

Corollary 3. *For any allowed value of the deformation parameter m ($m \in (0,1)$ for $g=1$ and $m \in \left(0, \frac{1}{2}\right)$ for $g=-1$) there is a discrete set $\{k_n\}_{n=1}^{\infty}$ of positive real numbers, increasingly ordered, such that $\phi_n = \chi_{m,k_n}^{(\pm)}\big|_{[0,\ell]}$ is a solution of the NLS equation on the interval $[0,\ell]$ with spectral parameters $\mu_n = k_n^2$ and n is the number of nodal domains. Furthermore, these are all the solutions of the NLS equation whose deformation parameter equals m.*

While this is mathematically sound, fixing the deformation parameter m is not a very useful approach in an applied setting. A more physical approach (and one that is useful when we consider star graphs) is to fix the L^2-norm $N_{n,\ell}^{(\pm)}(m) = \int_0^\ell \chi_{m,k_{n,\ell}^{(\pm)}(m)}^{(\pm)}(x)^2 dx$ of the solutions. The L^2-norm is a global measure for the strength of the nonlinearity. It has the physical meaning of an integrated intensity. In optical applications this is proportional to the total physical energy and for applications in Bose-Einstein condensates this is proportional to the number of particles.

By direct calculation (see [15]) we express the L^2-norms in terms of elliptic integrals (see Appendix A) as

$$\begin{aligned} N_{n,\ell}^{(+)}(m) &= \frac{8n^2}{\ell} K(m)\left[K(m) - E(1,m)\right] \\ N_{n,\ell}^{(-)}(m) &= \frac{8n^2(1-m)}{\ell} K(m)\left[\Pi(1,m,m) - K(m)\right], \end{aligned} \tag{15}$$

and use those to implicitly define the spectral curves in the form $k_{n,\ell}^{(\pm)}(N)$. The latter spectral curves are shown in Figure 4. The monotonicity of the spectral curves in this form follows from the monotonicity of $k_{n,\ell}^{(\pm)}(m)$ together with the monotonicity of $N_{n,\ell}^{(\pm)}(m)$. More precisely, one may check that $N_{n,\ell}^{(\pm)}(m)$ in (15) is an increasing function of m in the corresponding interval $m \in (0,1]$ for $g=1$ and $m \in \left(0, \frac{1}{2}\right)$ for $g=-1$. To verify this statement, observe that

1. $[K(m) - E(1,m)]$ and $\frac{1}{m}[\Pi(1,m,m) - K(m)]$ are increasing functions of m. This follows from their integral representations (see Appendix A). Explicitly, writing each expression as an integral, the corresponding integrands are positive and pointwise increasing functions of m.
2. $K(m)$ and $m(1-m)$ are also positive increasing functions of m in the relevant intervals.

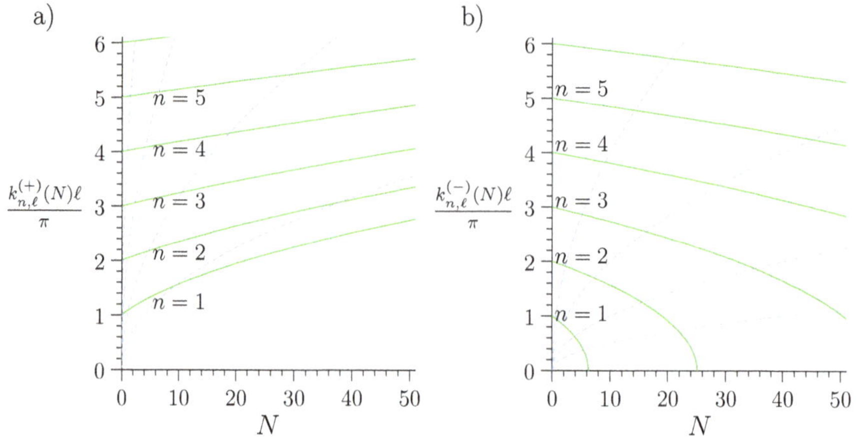

Figure 4. The spectral curves $k^{(\pm)}_{n,\ell}(N)$ (full green lines) in the repulsive (**a**) and attractive case (**b**). The n-th curve is obtained from the curve for $n = 1$ by scaling $k^{(\pm)}_{n,\ell}(N) = nk^{(\pm)}_{1,\ell}(N/n^2)$. The (blue) dashed lines indicate trajectories of the flow (17). The deformation parameter m is constant along the flow.

The inverse of $N^{(\pm)}_{n,\ell}(m)$ will be denoted $m^{(\pm)}_{n,\ell}(N)$. Combining the monotonicity of $k^{(\pm)}_{n,\ell}(m)$ and $N^{(\pm)}_{n,\ell}(m)$ one finds in the repulsive case that $k^{(+)}_{n,\ell}(N)$ is an increasing function of N defined for $N > 0$ while in the attractive case $k^{(-)}_{n,\ell}(N)$ is a decreasing function defined on $0 < N < N^{(-),\max}_{n,\ell}$ where

$$N^{(-),\max}_{n,\ell} = \frac{4n^2}{\ell} K\left(\frac{1}{2}\right)\left[\Pi\left(1, \frac{1}{2}, \frac{1}{2}\right) - K\left(\frac{1}{2}\right)\right]. \tag{16}$$

A characterization of the spectral curves $k^{(\pm)}_{n,\ell}(N)$ may be given as follows. We define the following flow in the k-N-plane (see Figure 4).

$$\Phi^{\tau}(m) = (N^{(\pm)}_{\tau,\ell}(m), k^{(\pm)}_{\tau,\ell}(m)), \tag{17}$$

where $N^{(\pm)}_{\tau,\ell}(m)$ and $k^{(\pm)}_{\tau,\ell}(m)$ are extensions of the expressions in Equations (15) and (14), replacing the integer valued n with the real flow parameter τ. Observe that $k^{(\pm)}_{\tau,\ell}(m)$ depends linearly on τ whereas, $N^{(\pm)}_{\tau,\ell}(m)$ is proportional to τ^2. This means that for each value of m, the corresponding flow line $\{\Phi^{\tau}(m)\}^{\infty}_{\tau=0}$ is of the form $k = \gamma\sqrt{N}$ (where γ depends on m). In particular, this implies that the spectral curves $k^{(\pm)}_{n,\ell}(N)$ are self-similar

$$k^{(\pm)}_{n,\ell}(N) = nk^{(\pm)}_{1,\ell}\left(\frac{N}{n^2}\right). \tag{18}$$

In addition, each flow line traverses the spectral curves $k^{(\pm)}_{n,\ell}(N)$ in the order given by the number of nodal domains n. This implies that the spectral curves never cross each other and remain properly ordered. We thus obtain the following second generalization of Sturm's oscillation theorem on the interval.

Proposition 3. *For $g = 1$ (repulsive case) let $N > 0$ and for $g = -1$ (attractive case) let $N \in (0, N^{(-),\max}_{1,\ell})$. Then there is a discrete set $\{k_n\}^{\infty}_{n=1}$ of positive real numbers, increasingly ordered such that*

$$\phi_n = \chi^{(\pm)}_{m^{(\pm)}_{n,\ell}(N),k_n}\Big|_{[0,\ell]}$$

is a solution of the NLS equation on the interval with a spectral parameter $\mu_n = k_n^2$ *and* L^2*-norm* $N = \int_0^\ell \phi_n(x)^2\, dx$ *and n is the number of nodal domains. Furthermore, these are all solutions whose* L^2*-norm equals N.*

3.2. Nonlinear Quantum Star Graphs

One may use the functions $\chi^{(\pm)}_{m,k}(x)$ defined in Equation (10) in order to reduce the problem of finding a solution of the NLS equation on a star graph to a (nonlinear) algebraic problem. By setting

$$\phi_e(x_e) = \sigma_e \chi^{(\pm)}_{m_e,k}(\ell_e - x_e) \tag{19}$$

where an overall sign $\sigma_e = \pm 1$ and the deformation parameter m_e remain unspecified (and allowed to take different values on different edges) one has a set of E functions that satisfy the NLS equation with spectral parameter $\mu = k^2$ on each edge and also satisfy the Dirichlet condition $\phi_e(\ell_e) = 0$ at the boundary vertices. Setting $\sigma_e = \operatorname{sgn}\left(\chi^{(\pm)}_{m_e,k}(\ell_e)\right)$ (unless $\chi^{(\pm)}_{m_e,k}(\ell_e) = 0$) the Kirchhoff matching conditions at the centre give a set of E independent nonlinear algebraic equations (see Equations (2) and (3)) for E continuous parameters $\{m_e\}$. If k is fixed there are typically discrete solutions for the parameters $\{m_e\}$. As k varies the solutions deform and form one-parameter families. Setting

$$N = \sum_{e=1}^{E} \int_0^{\ell_e} \phi_e(x_e)^2\, dx_e \tag{20}$$

each solution may be characterized by a pair (k, N) and as k is varied one naturally arrives at spectral curves in the k-N-plane, that may be expressed as $k(N)$ (or $N(k)$), as we have seen for the interval in the previous section. Nevertheless, the spectral curves of the star graph have a more intricate structure (see Figure 5). In non-linear algebraic equations one generally expects that solutions appear or disappear in bifurcations. For any particular example some numerical approach is needed to find the spectral curves. To do so, one first needs to have some approximate solution (either found by analytical approximation or by a numerical search in the parameter space). After that Newton-Raphson methods may be used to find the solution up to the desired numerical accuracy and the spectral curves are found by varying the spectral parameter slowly.

Figure 5 shows spectral curves that have been found numerically for a star graph with $E = 3$ and edge lengths $\ell_e = \sqrt{e}$ ($e = 1, 2, 3$). Most of the curves have been found starting from the corresponding spectrum of the linear problem ($g = 0$). Yet, one can see an additional curve that does not connect to the linear spectrum as $N \to 0$. This has originally been found in previous work [15] by coincidence, as the the numerical method jumped from one curve to another where they almost touch in the diagram.

We stress that in a numerical approach it is very hard to make sure that all solutions of interest are found, even if one restricts the search to a restricted region in parameter space. A full characterization of all solutions (such as given above for the nonlinear interval) will generally be elusive even for basic nonlinear quantum graphs. Theorems 1 and 2 and the related Corollaries 1 and 2 establish the existence of a large set of solutions inside the deep nonlinear regime. Each of these solutions may be used as a starting point for a numerical calculation of further solutions along the corresponding spectral curves.

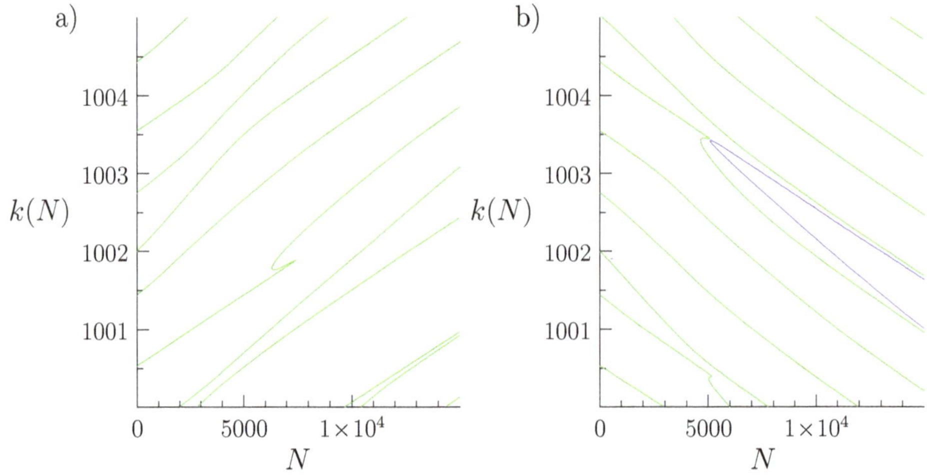

Figure 5. Spectral curves $k(N)$ for a nonlinear star graph with $E = 3$ edges and edge lengths $\ell_1 = 1$, $\ell_2 = \sqrt{2}$, and $\ell_3 = \sqrt{3}$ with repulsive (**a**) and attractive (**b**) nonlinear interaction. The spectral curves have been obtained by numerically solving the matching conditions using a Newton-Raphson method. For $N \to 0$ one obtains the spectrum of the corresponding linear star graph. Apart from one curve, all shown curves are connected to the linear spectrum this way. In the attractive case one spectral curve (shown in blue) is not connected to the linear spectrum. Such curves can sometimes be found by coincidence, e.g., if one is close to a bifurcation and numerical inaccuracy allows to jump from one solution branch to another (and this is indeed how we found it). In the repulsive case there is one spectral curve that has a sharp cusp. This indicates that there may be a bifurcation nearby that has additional solution branches that have not been found. In general it is a non-trivial numerical task to ensure that a diagram of spectral curves is complete. Here, completeness has not been attempted as the picture serves a mainly illustrative purpose.

3.3. Nodal Edge Counting and Central Dirichlet Solutions

It is interesting to consider the nodal structure along a spectral curve. Generically the wavefunction does not vanish at the centre and the nodal edge count structure (i.e., the vector $\mathbf{n}$) remains constant along the curve. The existence of central Dirichlet solutions implies that nodal points may move into (and through) the centre along a spectral curve (see also Theorem 2.9, [16]). At this instance the nodal edge count structure changes twice; first when the node hits the centre and then again when it has moved through. If $\mathbf{n}_0$ is the nodal edge count structure at a central Dirichlet solution, then generically the value of the function at the centre will change its sign along the spectral curve close to the central Dirichlet solution. If $\mathbf{n}_<$ and $\mathbf{n}_>$ are the nodal edge count structures close to the central Dirichlet solution then their entries differ at most by one $n_{>,e} - n_{<,e} = \pm 1$ and when the nodal point hits the centre one has $n_{0,e} = \min(n_{>,e}, n_{<,e})$. This is shown in more detail for a numerical example in Figure 6, where some central Dirichlet solutions are indicated on the spectral curves. The figure also shows the relevance of the central Dirichlet solutions for finding numerical solutions. The central Dirichlet solutions can be constructed directly using the machinery of the proof in the next section. From that one can then obtain a full spectral curve numerically by varying the parameters appropriately.

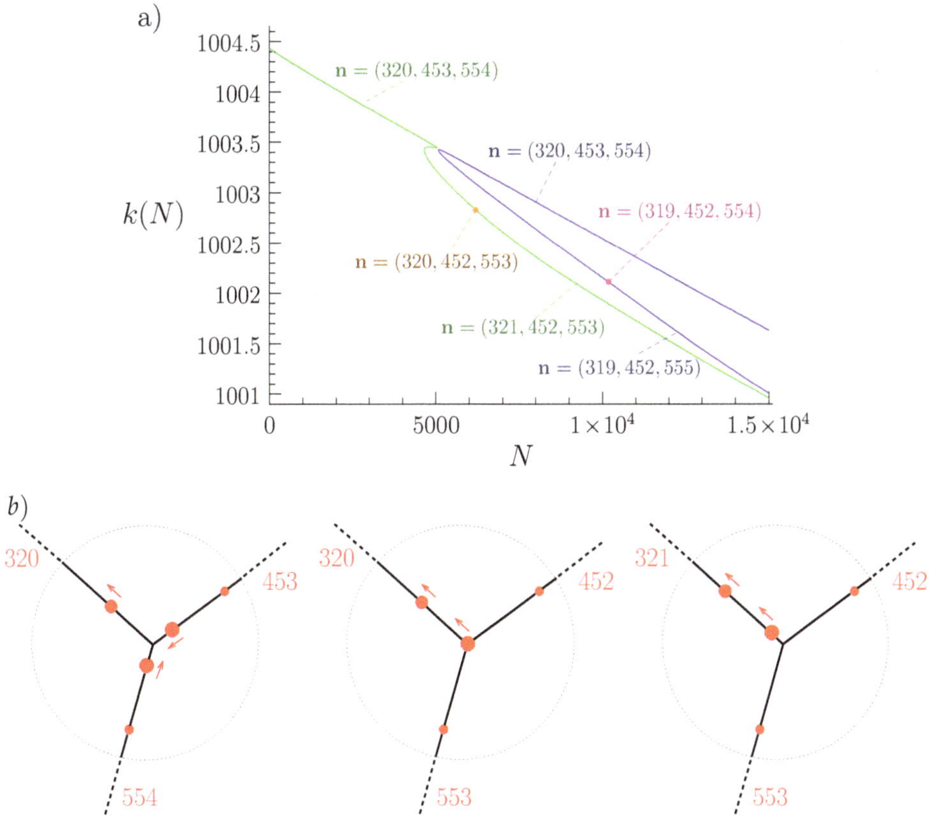

Figure 6. Upper panel (**a**): Two spectral curves (green and blue) of the star graph described in the caption of Figure 5 with attractive nonlinear interaction. The yellow and pink dots indicate positions that correspond to central Dirichlet solutions. The nodal edge structure **n** is indicated for each part of the curve. The latter is constant along spectral curves apart from jumps at the positions that correspond to central Dirichlet solutions. Lower panel (**b**): The three diagrams show how the nodal points move through the centre while N is increased through a central Dirichlet point on a spectral curve (the green curve in the upper panel). Only some nodal points close to the centre are shown. In the left diagram the three large dots are the closest to the centre and the arrows indicate how they move when N is increased. The numbers give the number of nodal domains on each edge. Increasing N further two nodal points on different edges merge at the centre as shown in the middle diagram. On the corresponding edges one nodal domain disappears. Further increasing N the nodal point moves from the centre into the remaining edge where the number of nodal domains is increased by one.

4. Proofs of Main Theorems

We prove the two theorems for repulsive and attractive interaction separately. The main construction is however the same. We start by describing the idea behind the construction and then turn to the actual proofs. Let $m^{(\pm)}_{1,\ell_e}(k)$ be the functions describing the deformation parameter of solutions on the interval of length ℓ with Dirichlet boundary conditions and a single nodal domain (they are given as the inverse of Equation (14); see also the lowest curve in Figure 3). Those functions are well-defined for $k > \frac{\pi}{\ell_e}$ in the repulsive case and for $k < \frac{\pi}{\ell_e}$ in the attractive case. Using these, we define

$$\phi_e := \sigma_e \, \chi^{(\pm)}_{m^{(\pm)}_{1,\ell_e}(k),k}\Big|_{[0,\ell_e]}, \tag{21}$$

for $e = 1, \ldots, E$, and where $\sigma_e = \pm 1$ are signs that will be specified later. These are of course just the solutions of the NLS with one nodal domain on the corresponding interval at spectral parameter $\mu = k^2$. In order to ensure that this function is well defined on all edges at given k we have to choose $k \in (\frac{\pi}{\ell_1}, \infty)$ in the repulsive case and $k \in (0, \frac{\pi}{\ell_E})$ in the attractive case (recall that we ordered the edge lengths by $\ell_1 < \ldots < \ell_E$).

As $\phi_e(0) = 0$ by construction, the set $\{\phi_e\}_{e=1}^E$ defines a continuous function on the graph including the centre for all allowed values of k. However, in general, these functions do not satisfy the remaining Kirchhoff condition $\sum_{e=1}^E \frac{d\phi_e}{dx_e}(0) = 0$. The idea of the proofs is the following. We consider $\sum_{e=1}^E \frac{d\phi_e}{dx_e}(0)$ as a function of k and need to show that it vanishes at some $k = k_0$. We find a particular set of signs $\{\sigma_e\}$ for which it is easy to show that $\sum_{e=1}^E \frac{d\phi_e}{dx_e}(0)$ changes sign as k is varied within its allowed range. Since this function is continuous in k it must vanish somewhere, which establishes the required central Dirichlet solution with exactly one nodal domain on each edge.

As we recognized above the role which the derivative of the solution plays in the proof, let us now directly calculate it.

$$\theta^{(\pm)}(m) := \frac{1}{\sqrt{2k^2}} \frac{d\chi^{(\pm)}_{m,k}}{dx}(0) = \begin{cases} \frac{\sqrt{m}}{1+m} & \text{for } g = 1, \\ \frac{\sqrt{m(1-m)}}{1-2m} & \text{for } g = -1. \end{cases} \tag{22}$$

In particular, expressing the derivative as a function of m and k and multiplying by a factor $\frac{1}{\sqrt{2k^2}}$, we see that the resulting function $\theta^{(\pm)}(m)$ does not depend explicitly on k, but only via the deformation parameter, m.

Remark 1. *In the statement of the theorem we have assumed that all edge lengths are different and stated how this may be relaxed in a subsequent remark. We can explain this now in more detail. Assume that we have two edge lengths that coincide. Denote those edges by e_0, e_1 and follow the construction above. By choosing opposite sign for the two edges $\sigma_{e_0} = \sigma_{e_1}$ the contribution of the two edges to the sum of derivatives $\sum_{e=1}^E \frac{d\phi_e}{dx_e}(0)$ cancels exactly for all allowed values of k, that is $\frac{d\phi_{e_0}}{dx_{e_0}}(0) + \frac{d\phi_{e_1}}{dx_{e_1}}(0) = 0$. One may then focus on the subgraph where the two edges are deleted and continue to construct a solution on the subgraph.*
One may also start with a graph with different edge lengths. If one has found any regular central Dirichlet solution on the graph one may add as many pairs of edges of the same length and find a regular non-Dirichlet solution on the larger graph following the above construction.

4.1. The Repulsive Case $g = 1$:

Proof of Theorem 1. Using the construction defined above we have to establish that there is a choice for the signs $\boldsymbol{\sigma} = (\sigma_1, \ldots, \sigma_E)$ and value for the spectral parameter $k = \sqrt{\mu}$ such that the Kirchhoff condition $\sum_{e=1}^\infty \frac{d\phi_e}{dx_e}(0) = 0$ is satisfied. For $k \in (\frac{\pi}{\ell_1}, \infty)$ let us define the function

$$f_\sigma(k) := \sum_{e=1}^E \sigma_e \theta^{(+)}\left(m^{(+)}_{1,\ell_e}(k)\right) = \sum_e \sigma_e \frac{\sqrt{m^{(+)}_{1,\ell_e}(k)}}{1 + m^{(+)}_{1,\ell_e(k)}} \tag{23}$$

where $\theta^{(+)}(m)$ was defined in Equation (22) and $m^{(+)}_{1,\ell_e}(k)$ is the inverse of

$$k^{(+)}_{1,\ell_e}(m) = \frac{2}{\ell_e}\sqrt{1+m}K(m) \tag{24}$$

as defined in Equation (14) (setting $n = 1$ for one nodal domain). The Kirchhoff condition is equivalent to the condition $f_\sigma(k) = 0$ for some $k > \frac{\pi}{\ell_1}$.

To continue the proof, we point out some monotonicity properties of $\theta^{(+)}(m)$ and $m^{(+)}_{1,\ell_e}(k)$. These properties may be easily verified by direct calculation using Equations (14) and (22). For $m \in (0,1)$ the functions $\theta^{(+)}(m)$ and $k^{(+)}_{1,\ell_e}(m)$ are strictly increasing and

$$\theta^{(+)}(0) = 0, \quad \theta^{(+)}(1) = \frac{1}{2}, \tag{25}$$

$$k^{(+)}_{1,\ell_e}(0) = \frac{\pi}{\ell_e}, \quad k^{(+)}_{1,\ell_e}(m) \underset{m \to 1}{\longrightarrow} \infty. \tag{26}$$

This implies that $\theta^{(+)}\left(m^{(+)}_{1,\ell_e}(k)\right)$ is strictly increasing for $k \in (\frac{\pi}{\ell_e}, \infty)$ and

$$\theta^{(+)}\left(m^{(+)}_{1,\ell_e}\left(\frac{\pi}{\ell_e}\right)\right) = 0 \quad , \quad \theta^{(+)}\left(m^{(+)}_{1,\ell_e}(k)\right) \underset{k \to \infty}{\longrightarrow} \frac{1}{2}. \tag{27}$$

If $E \geq 3$ is odd we choose the signs $\boldsymbol{\sigma} = (\sigma_1, \ldots, \sigma_E)$ to satisfy the following conditions

$$\sigma_1 = -1 \quad \text{and} \quad \sum_{e=2}^{E} \sigma_e = 0 \tag{28}$$

and

$$f_\sigma\left(\frac{\pi}{\ell_1}\right) = \sum_{e=1}^{E} \sigma_e \theta^{(+)}\left(m^{(+)}_{1,\ell_e}\left(\frac{\pi}{\ell_1}\right)\right) > 0\,. \tag{29}$$

Such a choice of signs is always possible as $\theta^{(+)}\left(m^{(+)}_{1,\ell_e}\left(\frac{\pi}{\ell_1}\right)\right) > 0$ for all $e > 1$. We then get by Equations (27) and (28) that $\lim_{k\to\infty} f_\sigma(k) = \frac{1}{2}\sum_{e=1}^{E} \sigma_e = -\frac{1}{2}$. By continuity there exists $k_0 \in (\frac{\pi}{\ell_1}, \infty)$ such that $f_\sigma(k_0) = 0$ for the given choice of signs. This proves the theorem for odd E.

For an even number of edges $E = 2M$ ($M \geq 2$) one needs to do a little bit more work. In this case, there are two strategies for choosing signs, $\boldsymbol{\sigma} = (\sigma_1, \ldots, \sigma_E)$, and showing that $f_\sigma(k)$ vanishes for some k.

1. One may choose more negative signs than positive signs so that $\sum_e \sigma_e < 0$. Then $\lim_{k\to\infty} f_\sigma(k) = \frac{1}{2}\sum_{e=1}^{E} \sigma_e$ is trivially negative. The difficulty here is in showing that such a choice is consistent with $f_\sigma\left(\frac{\pi}{\ell_1}\right) > 0$. This generally leads to some conditions which the edge lengths should satisfy.
2. One may choose as many positive as negative signs, which makes it easier to satisfy $f_\sigma\left(\frac{\pi}{\ell_1}\right) > 0$ (i.e., the conditions on the edge lengths are less restrictive). Yet, the difficulty here lies in $\lim_{k\to\infty} f_\sigma(k) = 0$, which means that one needs to show that this limit is approached from the negative side (i.e., find the conditions on the edge lengths which ensures this).

These two strategies give some indication on how our proof may be generalized beyond the stated length restrictions. Moreover, they also give a practical instruction for how one may search for further solutions numerically.

We continue the proof by following the second strategy and setting

$$\sigma_e = \begin{cases} 1 & \text{for } e = 1 \text{ and } e \geq M+2, \\ -1 & \text{for } 2 \leq e \leq M+1 \end{cases} \tag{30}$$

so that $\sum_{e=1}^{E} \sigma_e = 0$. One then has $\lim_{k\to\infty} f_\sigma(k) = 0$ and we will show that the leading term in the (convergent) asymptotic expansion of $f_\sigma(k)$ for large k is negative. Using the known asymptotics [17] of the elliptic integral $K(m)$ as $m = 1 - \delta m$ goes to one (or $\delta m \to 0$)

$$K(1-\delta m) = -\frac{1}{2}\log(\delta m) + 2\log(2) + O\left(\delta m \, \log(\delta m)\right) \tag{31}$$

one may invert Equation (24) asymptotically for large k as

$$1 - m^{(\pm)}_{1,\ell_e}(k) = 16e^{-\frac{k\ell_e}{\sqrt{2}}} + O\left(ke^{-\sqrt{2}k\ell_e}\right) \tag{32}$$

and, thus

$$\begin{aligned}\frac{\sqrt{m^{(\pm)}_{1,\ell_e}(k)}}{1+m^{(\pm)}_{1,\ell_e}(k)} =& \frac{\sqrt{1-\left(1-m^{(\pm)}_{1,\ell_e}(k)\right)}}{2-\left(1-m^{(\pm)}_{1,\ell_e}(k)\right)} \\ =& \frac{1}{2} - \frac{1}{16}\left(1-m^{(\pm)}_{1,\ell_e}(k)\right)^2 + O\left(\left(1-m^{(\pm)}_{1,\ell_e}(k)\right)^3\right) \\ =& \frac{1}{2} - 16e^{-\sqrt{2}k\ell_e} + O\left(ke^{-3\frac{k\ell_e}{\sqrt{2}}}\right).\end{aligned} \tag{33}$$

This directly leads to the asymptotic expansion

$$f_\sigma(k) = -\sum_{e=1}^{E}\sigma_e 16e^{-\sqrt{2}k\ell_e} + O\left(ke^{-3\frac{k\ell_1}{\sqrt{2}}}\right) = -16e^{-\sqrt{2}k\ell_1}\left(1+\sum_{e=2}^{E}\sigma_e e^{-\sqrt{2}k(\ell_e-\ell_1)}\right) + O\left(ke^{-3\frac{k\ell_1}{\sqrt{2}}}\right) \tag{34}$$

which is negative for sufficiently large k because ℓ_1 is the shortest edge length.

It is left to show $f_\sigma\left(\frac{\pi}{\ell_1}\right) > 0$.

For this let us write $m^{(+)}_{1,\ell_e}(k) = m^{(+)}_{1,1}(k\ell_e)$ for each term. As $\theta^{(+)}\left(m^{(+)}_{1,1}(\pi)\right) = 0$ the condition $f_\sigma\left(\frac{\pi}{\ell_1}\right) > 0$ is equivalent to

$$\sum_{e=2}^{M+1}\theta^{(+)}\left(m^{(+)}_{1,1}\left(\frac{\pi\ell_e}{\ell_1}\right)\right) < \sum_{e=M+2}^{E}\theta^{(+)}\left(m^{(+)}_{1,1}\left(\frac{\pi\ell_e}{\ell_1}\right)\right), \tag{35}$$

using our choice of the signs, Equation (30). Since $m^{(+)}_{1,1}$ and $\theta^{(+)}$ are increasing functions and $\ell_1 < \ldots < \ell_E$, condition (35) is certainly satisfied if

$$\frac{\theta^{(+)}\left(m^{(+)}_{1,1}\left(\frac{\pi\ell_{M+2}}{\ell_1}\right)\right)}{\theta^{(+)}\left(m^{(+)}_{1,1}\left(\frac{\pi\ell_{M+1}}{\ell_1}\right)\right)} > \frac{M}{M-1}. \tag{36}$$

The condition (36) restricts the three edge lengths ℓ_1, ℓ_{M+1} and ℓ_{M+2} and it is equivalent to the condition (5) stated in the theorem. Indeed, this is trivial for the right-hand side where $\frac{M}{M-1} = \frac{E}{E-2}$. For the left-hand side note that Equation (6) in Theorem 1 identifies $m_+ = m^{(+)}_{1,1}\left(\frac{\pi\ell_{M+2}}{\ell_1}\right)$ and $m_- = m^{(+)}_{1,1}\left(\frac{\pi\ell_{M+1}}{\ell_1}\right)$ such that the left-hand-side of the stated condition (5) in the theorem and the left-hand side of Equation (36) are identical when written out explicitly. □

4.2. The Attractive Case $g = -1$:

Proof of Theorem 2. In the attractive case we can start similarly to the previous proof by rewriting the Kirchhoff condition on the sum of derivatives as $f_\sigma(k) = 0$ for some $k \in (0, \frac{\pi}{\ell_E})$ where

$$f_\sigma(k) = k^2\sum_{e=1}^{E}\sigma_e\theta^{(-)}\left(m^{(-)}_{1,\ell_e}(k)\right) = \sum_{e}^{E}\sigma_e\frac{4\sqrt{m^{(-)}_{1,\ell_e}(k)(1-m^{(-)}_{1,\ell_e}(k))}K\left(m^{(-)}_{1,\ell_e}(k)\right)^2}{\ell_e^2}. \tag{37}$$

The additional factor k^2 is irrelevant for satisfying the condition but allows us to extend the definition of the function to $k = 0$ (where $k^2 \sim 1 - 2m^{(-)}_{1,\ell}(k)$). Noting that $m^{(-)}_{1,\ell_e}(k)$ is a decreasing function for $k \in (0, \frac{\pi}{\ell_e})$ with $m^{(-)}_{1,\ell_e}(0) = \frac{1}{2}$ and $m^{(-)}_{1,\ell_e}\left(\frac{\pi}{\ell_e}\right) = 0$ and $K(m)$ is increasing with m we get that the function

$$k^2\theta^{(-)}\left(m^{(-)}_{1,\ell_e}(k)\right) = \frac{4\sqrt{m^{(-)}_{1,\ell_e}(k)(1 - m^{(-)}_{1,\ell_e}(k))}K\left(m^{(-)}_{1,\ell_e}(k)\right)^2}{\ell_e^2}$$

is a decreasing function for $k \in (0, \frac{\pi}{\ell_e})$ and

$$\lim_{k\to 0} k^2\theta^{(-)}\left(m^{(-)}_{1,\ell_e}(k)\right) = \frac{2K(\frac{1}{2})^2}{\ell_2^2} \quad , \quad \left(\frac{\pi}{\ell_e}\right)^2\theta^{(-)}\left(m^{(-)}_{1,\ell_e}\left(\frac{\pi}{\ell_e}\right)\right) = 0.$$

Altogether this implies that

$$f_\sigma(0) = 2K\left(\frac{1}{2}\right)^2 \sum_{e=1}^{E} \frac{\sigma_e}{\ell_e^2} \tag{38}$$

and

$$f_\sigma\left(\frac{\pi}{\ell_E}\right) = \sum_{e=1}^{E-1} \sigma_e \frac{4\sqrt{m^{(-)}_{1,\ell_e}\left(\frac{\pi}{\ell_E}\right)\left(1 - m^{(-)}_{1,\ell_e}\left(\frac{\pi}{\ell_E}\right)\right)}K\left(m^{(-)}_{1,\ell_e}\left(\frac{\pi}{\ell_E}\right)\right)^2}{\ell_e^2} \tag{39}$$

Now let us assume that the two conditions (8) stated in Theorem 2 are satisfied and let us choose (for $M < E/2$ as is given in the condition of the theorem)

$$\sigma_e = \begin{cases} 1 & \text{for } e \leq M, \\ -1 & \text{for } e \geq M+1. \end{cases} \tag{40}$$

Then

$$f_\sigma(0) = 2K\left(\frac{1}{2}\right)^2 \left[\sum_{e=1}^{M} \frac{1}{\ell_e^2} - \sum_{e=M+1}^{E} \frac{1}{\ell_e^2}\right] \tag{41}$$

and the right inequality of (8) directly implies that $f_\sigma(0) < 0$.

In order to prove the existence of the solution stated in Theorem 2, it is left to show that $f_\sigma\left(\frac{\pi}{\ell_E}\right) > 0$, which would imply that f_σ vanishes for some $k \in (0, \frac{\pi}{\ell_E})$. Using our choice of signs and the identity $m^{(-)}_{1,\ell_e}(k) = m^{(-)}_{1,1}(k\ell_e)$ we may rewrite Equation (39) as

$$\begin{aligned} f_\sigma\left(\frac{\pi}{\ell_E}\right) = &\sum_{e=1}^{M} \frac{4\sqrt{m^{(-)}_{1,1}\left(\frac{\pi\ell_e}{\ell_E}\right)\left(1 - m^{(-)}_{1,1}\left(\frac{\pi\ell_e}{\ell_E}\right)\right)}K\left(m^{(-)}_{1,1}\left(\frac{\pi\ell_e}{\ell_E}\right)\right)^2}{\ell_e^2} \\ &- \sum_{e=M+1}^{E-1} \frac{4\sqrt{m^{(-)}_{1,1}\left(\frac{\pi\ell_e}{\ell_E}\right)\left(1 - m^{(-)}_{1,1}\left(\frac{\pi\ell_e}{\ell_E}\right)\right)}K\left(m^{(-)}_{1,1}\left(\frac{\pi\ell_e}{\ell_E}\right)\right)^2}{\ell_e^2}. \end{aligned} \tag{42}$$

As $\sqrt{m(1-m)}K(m)^2$ is an increasing function for $m \in (0, \frac{1}{2})$ and $m^{(-)}_{1,1}(k)$ is a decreasing function of its argument the left inequality in Equation (8) implies

$$\sqrt{m_-(1-m_-)}K(m_-)^2 \sum_{e=M+1}^{E-1} \frac{1}{\ell_e^2} < \sqrt{m_+(1-m_+)}K(m_+)^2 \sum_{e=1}^{M} \frac{1}{\ell_e^2} \tag{43}$$

where $m_+ = m_{1,1}^{(-)}\left(\frac{\pi \ell_M}{\ell_1}\right)$ and $m_- = m_{1,1}^{(-)}\left(\frac{\pi \ell_{M+1}}{\ell_1}\right)$. The same monotonicity argument implies that the negative contributions in Equation (42) are smaller than the left-hand side of inequality Equation (43) and that the positive contributions in Equation (42) are larger than the right-hand side of inequality Equation (43). Thus Equation (43) implies $f_\sigma\left(\frac{\pi}{\ell_E}\right) > 0$, as required. □

5. Conclusions

We have established the existence of solutions of the stationary nonlinear Schrödinger equation on metric star graphs with a nodal point at the centre. The existence is subject to certain conditions on the edge lengths that can be satisfied for any numbers of edges $E \geq 3$. We stress that some of these solutions are deep in the nonlinear regime where finding any solutions is quite non-trivial. Let us elaborate on that. The non-linear solutions come in one-parameter families, and a possible way to track those families (or curves) is to start from the solutions of the corresponding linear Schrödinger equation. Indeed, since the linear solutions are good approximations for the nonlinear solutions with low intensities they may be used as starting points for finding nonlinear solutions numerically. By slowly changing parameters one may then find some spectral curves that extend into the deep nonlinear regime. However, there may be many solutions on spectral curves that do not extend to arbitrary small intensities and these are are much harder to find numerically. By focusing on solutions which vanish at the centre our work shows how to construct such solutions. These may then be used in a numerical approach to give a more complete picture of spectral curves.

The solutions that we construct are characterised by their nodal count structure. The nodal structure on a spectral curve is constant as long as the corresponding solutions do not vanish at the centre. Our approach thus constructs the solutions where the nodal structure changes along the corresponding spectral curve. In this way we have made some progress in characterizing general solutions on star graphs in terms of their nodal structure.

Many open questions remain. The main one being whether all spectral curves of the NLS equation on a star graph may be found just by combining the linear solutions with the non-linear solutions which vanish at the centre. If not, how many other spectral curves remain and how can they be characterized? Numerically we found that apart from the ground state spectral curve it is generic for a spectral curve to have at least one point where the corresponding solution vanishes at the centre. Of course this leaves open how many spectral curves there are where the corresponding solutions never vanishes at the centre. Another interesting line of future research may be to extend some of our results to tree graphs.

Author Contributions: Conceptualization, R.B. and S.G.; methodology, R.B., S.G. and A.J.K.; software, S.G. and A.J.K.; validation, R.B., S.G. and A.J.K.; formal analysis, R.B., S.G. and A.J.K.; writing–original draft preparation, A.J.K. and S.G.; writing–review and editing, R.B. and S.G.; visualization, R.B. and S.G.; supervision, R.B. and S.G.; project administration, R.B. and S.G.; funding acquisition, R.B. and S.G.

Funding: R.B. and A.J.K. were supported by ISF (Grant No. 494/14). SG was supported by the Joan and Reginald Cohen Foundation.

Acknowledgments: S.G. would like to thank the Technion for hospitality and the Joan and Reginald Coleman-Cohen Fund for support.

Conflicts of Interest: The authors declare no conflict of interest.

Appendix A. Elliptic Integrals and Jacobi Elliptic Functions

We use the following definitions for elliptic integrals (the Jacobi form)

$$F(x|m) := \int_0^x \frac{1}{\sqrt{1-u^2}\sqrt{1-m\,u^2}}du \tag{A1a}$$

$$K(m) := F(1|m) \tag{A1b}$$

$$E(x|m) := \int_0^x \frac{\sqrt{1-m\,u^2}}{\sqrt{1-u^2}}du \tag{A1c}$$

$$\Pi(x|a,m) := \int_0^x \frac{1}{\sqrt{1-u^2}\sqrt{1-m\,u^2}(1-a\,u^2)}du \tag{A1d}$$

where $0 \le x \le 1$, $m \le 1$ and $a \le 1$. Note that our definition allows m and a to be negative.

The notation in the literature is far from being uniform. Our choice seems the most concise for the present context and it is usually straight-forward to translate our definitions into the ones of any standard reference on special functions. For instance, the *NIST Handbook of Mathematical Functions* [17] defines the three elliptical integrals $F(\phi, k)$, $E(\phi, k)$ and $\Pi(\phi, \alpha, k)$ by setting $x = \sin(\phi)$, $m = k^2$, and $a = \alpha^2$ in our definitions above.

Jacobi's Elliptic function $\mathrm{sn}(x, m)$, the elliptic sine, is defined as the inverse of $F(u|m)$

$$u = \mathrm{sn}(x,m) \qquad \Leftrightarrow \qquad x = F(u|m)\,. \tag{A2}$$

This defines $\mathrm{sn}(x, m)$ for $x \in [0, K(m)]$ which can straight-forwardly be extended to a periodic function with period $4K(m)$ by requiring $\mathrm{sn}(K(m) + x, m) = \mathrm{sn}(K(m) - x, m)$, $\mathrm{sn}(-x, m) = -\mathrm{sn}(x, m)$ and $\mathrm{sn}(x + 4K(m), m) = \mathrm{sn}(x, m)$. The corresponding elliptic cosine $\mathrm{cn}(x, m)$ is obtained by requiring that it is a continuous function satisfying

$$\mathrm{cn}^2(x,m) + \mathrm{sn}^2(x,m) = 1 \tag{A3}$$

such that $\mathrm{cn}(0, m) = 1$. It is useful to also define the non-negative function

$$\mathrm{dn}(x,m) := \sqrt{1 - m\,\mathrm{sn}^2(x,m)}. \tag{A4}$$

At $m = 0$ and $m = 1$ the elliptic functions can be expressed as

$$\mathrm{sn}(x,0) = \sin x, \qquad \mathrm{sn}(x,1) = \tanh x, \tag{A5a}$$

$$\mathrm{cn}(x,0) = \cos x, \qquad \mathrm{cn}(x,1) = \cosh^{-1} x, \tag{A5b}$$

$$\mathrm{dn}(x,0) = 1, \qquad \mathrm{dn}(x,1) = \cosh^{-1} x\,. \tag{A5c}$$

Derivatives of elliptic functions can be expressed in terms of elliptic functions

$$\frac{d}{dx}\mathrm{sn}(x,m) = \mathrm{cn}(x,m)\mathrm{dn}(x,m), \tag{A6a}$$

$$\frac{d}{dx}\mathrm{cn}(x,m) = -\,\mathrm{sn}(x,m)\mathrm{dn}(x,m), \tag{A6b}$$

$$\frac{d}{dx}\mathrm{dn}(x,m) = -\,m\,\mathrm{sn}(x,m)\mathrm{cn}(x,m)\,. \tag{A6c}$$

The first of these equations implies that $u = \mathrm{sn}(x, m)$ is a solution of the first order ordinary differential equation

$$\frac{du}{dx} = \sqrt{1-u^2}\sqrt{1-mu^2}\,. \tag{A7}$$

References

1. Sturm, C. Mémoire sur une classe d'équations à différences partielles. *J. Math. Pures Appl.* **1836**, *1*, 373–444.
2. Courant, R. Ein allgemeiner Satz zur Theorie der Eigenfunktionen selbstadjungierter Differentialausdrücke. *Nachr. Ges. Wiss. Göttingen Math Phys.* **1923**, *K*1, 81–84.
3. Ancona, A.; Helffer, B.; Hoffmann-Ostenhof, T. Nodal domain theorems à la Courant. *Doc. Math.* **2004**, *9*, 283–299.
4. Pleijel, A. Remarks on courant's nodal line theorem. *Commun. Pure Appl. Math.* **1956**, *9*, 543–550. [CrossRef]
5. Gnutzmann, S.; Smilansky, U.; Weber, J. Nodal counting on quantum graphs. *Waves Random Med.* **2004**, *14*, S61–S73. [CrossRef]
6. Alon, L.; Band, R.; Berkolaiko, G. Nodal statistics on quantum graphs. *Commun. Math. Phys.* **2018**, *362*, 909–948. [CrossRef]
7. Band, R.; Berkolaiko, G.; Weyand, T. Anomalous nodal count and singularities in the dispersion relation of honeycomb graphs. *J. Math. Phys.* **2015**, *56*, 122111. [CrossRef]
8. Pokornyĭ, Y.V.; Pryadiev, V.L.; Al'-Obeĭd, A. On the oscillation of the spectrum of a boundary value problem on a graph. *Mat. Zametki* **1996**, *60*, 468–470. [CrossRef]
9. Schapotschnikow, P. Eigenvalue and nodal properties on quantum graph trees. *Waves Random Complex Med.* **2006**, *16*, 167–178. [CrossRef]
10. Band, R. The nodal count $\{0, 1, 2, 3, \dots\}$ implies the graph is a tree. *Philos. Trans. R. Soc. Lond. A* **2014**, *372*, 20120504. [CrossRef]
11. Berkolaiko, G. A lower bound for nodal count on discrete and metric graphs. *Commun. Math. Phys.* **2007**, *278*, 803–819. [CrossRef]
12. Band, R.; Berkolaiko, G.; Raz, H.; Smilansky, U. The number of nodal domains on quantum graphs as a stability index of graph partitions. *Commun. Math. Phys.* **2012**, *311*, 815–838. [CrossRef]
13. Berkolaiko, G.; Weyand, T. Stability of eigenvalues of quantum graphs with respect to magnetic perturbation and the nodal count of the eigenfunctions. *Philos. Trans. R. Soc. A* **2013**, *372*. [CrossRef]
14. Gnutzmann, S.; Waltner, D. Stationary waves on nonlinear quantum graphs: General framework and canonical perturbation theory. *Phys. Rev.* **2016**. [CrossRef] [PubMed]
15. Gnutzmann, S.; Waltner, D. Stationary waves on nonlinear quantum graphs: II. application of canonical perturbation theory in basic graph structures. *Phys. Rev. E* **2016**, *94*. [CrossRef] [PubMed]
16. Band, R.; Krueger, A.J. Nonlinear Sturm oscillation: From the interval to a star. In *Mathematical Problems in Quantum Physics*; American Mathematical Society: Providence, RI, USA, 2018; Volume 717, pp. 129–154.
17. Olver, F.W.J.; Lozier, D.W.; Boisvert, R.F.; Clark, C.W. *The NIST Handbook of Mathematical Functions*; Cambridge Univ. Press: Cambridge, UK, 2010.

symmetry

MDPI

Article

Soliton and Breather Splitting on Star Graphs from Tricrystal Josephson Junctions

Hadi Susanto [1], Natanael Karjanto [2,*], Zulkarnain [3], Toto Nusantara [4] and Taufiq Widjanarko [5]

1 Department of Mathematical Sciences, University of Essex, Colchester CO4 3SQ, UK; hsusanto@essex.ac.uk
2 Department of Mathematics, University College, Sungkyunkwan University, Natural Science Campus, 2066 Seobu-ro, Jangan-gu, Suwon 16419, Gyeonggi-do, Korea
3 Department of Mathematics, Faculty of Mathematics and Natural Sciences, Universitas Riau, Kampus Bina Widya KM 12.5, Simpang Baru, Tampan, Kota Pekanbaru, Riau 28293, Indonesia; zulqr27@gmail.com
4 Department of Mathematics, Faculty of Mathematics and Natural Sciences, Jalan Semarang 5, Malang 65145, Indonesia; toto.nusantara.fmipa@um.ac.id
5 Manufacturing Metrology Team, Faculty of Engineering, The University of Nottingham, Advanced Manufacturing Building, Jubilee Campus, Wollaton Road, Nottingham NG8 1BB, UK; Taufiq.Widjanarko@nottingham.ac.uk
* Correspondence: natanael@skku.edu

Received: 9 January 2019; Accepted: 12 February 2019; Published: 20 February 2019

Abstract: We consider the interactions of traveling localized wave solutions with a vertex in a star graph domain that describes multiple Josephson junctions with a common/branch point (i.e., tricrystal junctions). The system is modeled by the sine-Gordon equation. The vertex is represented by boundary conditions that are determined by the continuity of the magnetic field and vanishing total fluxes. When one considers small-amplitude breather solutions, the system can be reduced into the nonlinear Schrödinger equation posed on a star graph. Using the equation, we show that a high-velocity incoming soliton is split into a transmitted component and a reflected one. The transmission is shown to be in good agreement with the transmission rate of plane waves in the linear Schrödinger equation on the same graph (i.e., a quantum graph). In the context of the sine-Gordon equation, small-amplitude breathers show similar qualitative behaviors, while large-amplitude ones produce complex dynamics.

Keywords: soliton; breather; sine-Gordon equation; Schrödinger equation; star graph; quantum graph

1. Introduction

A quantum graph is a metric graph, i.e., a network-shaped structure of vertices connected by edges, with a Schrödinger-like operator suitably defined on functions that are supported on the edges. It arises as a model for wave propagations in a system similar to a thin neighborhood of a graph. Pauling [1] was most likely the pioneer of the research subject when he modeled free electrons in organic molecules. In his model, he approximated the atoms as vertices while the electrons form bonds that fix a frame in the shape of the molecule on which the free electrons are confined. The term 'quantum graph' itself may be a shortening of the title of a paper by Kottos and Smilansky [2]. See, e.g., [3] for an elementary introduction to quantum graphs, where some basic tools in the spectral theory of the Schrödinger operator on metric graphs are discussed.

Quantum graphs have been used to describe a variety of mathematical concepts as well as physical problems and applications. A review of quantum graphs with applications in theoretical physics is provided by Gnutzmann and Smilansky [4]. For a comprehensive introduction and survey of the current state of research on quantum graphs and their applications, the reader is encouraged to consult a mathematically oriented book by Berkolaiko and Kuchment [5]. See also an introduction and a brief survey to quantum graphs by Kuchment [6].

The study of nonlinear counterparts of quantum graphs, where the linear wave equations are replaced by nonlinear ones, has been growing due to their potential of becoming a paradigm model for topological effects in nonlinear wave propagation (see [7] for a recent review). Because of the nonlinearity, soliton solutions exist. However, a unique 'trapped soliton' state, which is admitted by the cubic focusing nonlinear Schrödinger (NLS) equation on the star graph with Kirchhoff conditions at the vertex, is not the ground state [8]. This is remarkably different from the NLS equation on the line. The existence and behavior of trapped solitons with a δ-interaction at the vertex are considered by Adami et al. [9]. A generalized NLS equation with power nonlinearity on star graphs has also been investigated in various reports [10–12]. The existence of ground states of the same equation on several types of star metric graphs has been considered in [13]. Bifurcations of stationary solutions in various other simple topologies have also been studied, such as in tadpole graphs consisting of a half-line joined to a loop at a single vertex [14], dumbbell-shaped metric graphs [15,16], bowtie graphs [16], and double-bridge graphs [17].

In addition to stationary solutions, the interaction of a moving soliton with the vertex is also intriguing. Soliton scattering in the NLS equation on a star graph with a repulsive δ-, δ'-function potential, and the free Kirchoff condition at the vertex is studied by Adami et al. [18], who extended the work of Holmer et al. [19]. Adami et al. showed that a soliton will split into a transmitted soliton, a reflected one, and some radiation upon collision with the vertex. Soliton dynamics in star graphs with 'integrable' vertex conditions that preserve the solution norm was studied in [20].

In this work, we consider a star graph that models a tricrystal Josephson junction, see Figure 1. A Josephson junction is a quantum mechanical structure that is made of two superconducting electrodes separated by a thin barrier. Three semi-infinite junctions with the ends meeting at a common point form a tricrystal junction. The vertex conditions were likely first derived in [21,22], where the structure was proposed as a logic gate device. In recent work, tricrystal Josephson junctions were fabricated as a probe of the order parameter symmetry of high-temperature superconductors [23–26]. In particular, tetracrystal junctions (i.e., star graphs with four arms) were also constructed and studied experimentally, see, e.g., Section IV.C of [26,27].

The study of soliton solutions in tricrystal junctions has only been done for topological solitons, i.e., kinks, especially when they are static [25,28–30]. This type of solitons is also called 'fluxons' because they carry integer quanta of electromagnetic flux. The dynamics of moving vortices in tricrystal junctions was discussed in [21,22,31,32]. Here, instead, we consider for the first time the dynamics of non-topological solitons, i.e., breathers (see Figure 1). In the case of small-amplitude breather solutions, the governing equation, which is the sine-Gordon (sG) equation, can be reduced into the NLS equation with vertex conditions different from those considered in previous works. Soliton dynamics within that approximation will be discussed as well.

The paper is structured as follows. In Section 2, we discuss the governing equation of tricrystal Josephson junctions. Under an assumption of small-amplitude solutions, we will also derive the NLS equation with vertex conditions. In Section 3, the scattering of linear plane waves, soliton and breather solutions will be discussed. Numerical simulations will be presented in the same section describing the scattering processes when nonlinearity is present. In Section 4, we consider the nonlinear scattering of sG breathers numerically. We present two different typical cases corresponding to small- and large-amplitudes with slow- and fast-incoming velocities. The conclusion of the paper is in Section 5.

2. Governing Equations

The phase difference of wave functions along each Josephson junction is described by the sG equation

$$u_{xx}^{(j)} - u_{tt}^{(j)} = \sin u^{(j)}, \quad (1)$$

where upper indices $j = 1, 2, 3$, label the different branches of the system and the subscripts indicate derivatives with respect to the variables. The direction of the x-axes follows the sketch in Figure 1.

At the meeting point between the three branches, i.e., $x = 0$, we have the boundary conditions [28,31]

$$u^{(1)} = u^{(2)} + u^{(3)}, \qquad u_x^{(1)} = u_x^{(2)} = u_x^{(3)}. \tag{2}$$

The first equation comes from the physical property that the magnetic flux through an infinitesimally small contour encircling the origin must vanish, i.e., the total change of the gauge-invariant phase difference is zero. The second equation means that the field, which is proportional to the slope of the phase difference, is continuous at the origin.

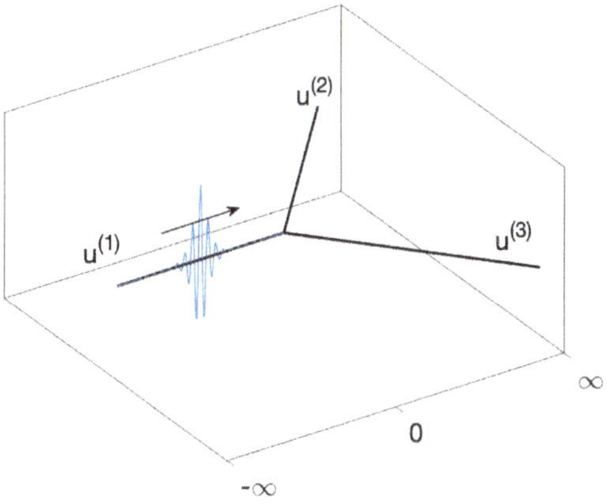

Figure 1. A schematic diagram of the system, showing a breather traveling from $x \ll 0$ towards $x = 0$. The convention of the spatial direction used herein is indicated.

Far away from the origin $x \ll 0$, the sG Equation (1) admits two types of fundamental solitons: a kink, that is in the form of

$$u(x,t) = 4\tan^{-1}\left\{\exp[-\gamma(x - vt - x_0)]\right\}, \qquad x_0 \in \mathbb{R} \tag{3}$$

and a breather

$$u(x,t) = 4\tan^{-1}\left\{\tan\theta \sin\left[\gamma\cos(\theta)(t - vx - t_0)\right]\right] \operatorname{sech}\left[\gamma\sin(\theta)(x - vt - x_0)\right]\right\}, \qquad x_0 \in \mathbb{R} \tag{4}$$

where $\gamma = 1/\sqrt{1-v^2}$ is the Lorentz-type contraction of a moving excitation. Both represent a topological and non-topological soliton moving with velocity v. The breather (4) oscillates in time with frequency $\gamma\cos\theta$. All the previous work on solitons in tricrystal junctions dealt with the first class of solutions (see the related references mentioned in Section 2). Here, we will concentrate on breather dynamics.

To study the dynamics of small-amplitude breathers (4), one can also use the standard multiple-scale expansion method. Writing

$$u^{(j)}(x,t) = \epsilon U^{(j)}(X,T) + \text{c.c.} + \text{higher-order terms},$$

where $X = \epsilon x$ and $T = \epsilon^2 t/2$ are the slow space and time variables, respectively, and c.c. is the complex conjugation of the preceding terms, we obtain the NLS equation (see, e.g., [33])

$$iU_T^{(j)} + U_{XX}^{(j)} + \frac{1}{2}|U^{(j)}|^2 U^{(j)} = 0. \tag{5}$$

The boundary conditions (2) then become

$$U^{(1)} = U^{(2)} + U^{(3)}, \qquad U_X^{(1)} = U_X^{(2)} = U_X^{(3)}. \tag{6}$$

Far away from the branch point $X = 0$, i.e., $X \to -\infty$, the NLS equation has a traveling bright soliton

$$U(X,T) = A\,\mathrm{sech}(A(X - vT - X_0)/2)\exp(i\phi - ivX/2 + i(A^2 - v^2)T/4), \tag{7}$$

with $A > 0$, X_0, ϕ and $v \in \mathbb{R}$. This soliton approximates the breather solution (4) for small $|\theta|$.

3. Scattering of NLS Solitons

Studying the scattering problem for non-topological solitary waves as sketched in Figure 1, we can consider a symmetric solution between the second and third branch, i.e., $u^{(2)} = u^{(3)}$ and $U^{(2)} = U^{(3)}$. Under the symmetry, the initial boundary value problems (1)–(2) and (5)–(6) become a single equation on the real line:

$$u_{xx} - u_{tt} = \sin u, \qquad x \in \mathbb{R}, \tag{8}$$

with $u(0^-) = 2u(0^+)$, $u_x(0^-) = u_x(0^+)$ and

$$iU_T + U_{XX} + \frac{1}{2}|U|^2U = 0, \qquad X \in \mathbb{R}, \tag{9}$$

with $U(0^-) = 2U(0^+)$, $U_X(0^-) = U_X(0^+)$.

3.1. Scattering in the Linear Problems

It is natural to consider first the scattering problem for plane waves in the linear Schrödinger equation obtained by omitting the cubic term in Equtaion (9). Incoming plane waves arriving from $X \to -\infty$ has the form $U(X,T) = e^{i(kX-\omega T)}$, with the amplitude normalized to 1, $\omega > 0$, and $k = \sqrt{\omega}$. The general solution of the scattering problem is

$$\psi(X) = \begin{cases} e^{i(kX-\omega T)} + \tilde{r}e^{i(-kX-\omega T)}, & X < 0, \\ \tilde{t}e^{i(kX-\omega T)}, & X > 0. \end{cases} \tag{10}$$

Here, $\tilde{r}$ and $\tilde{t}$ are reflection and transmission coefficients, respectively.

Substituting the solution (10) into the boundary conditions at $X = 0$ will give us

$$\tilde{r} = \frac{1}{3}, \qquad \tilde{t} = \frac{2}{3}. \tag{11}$$

It turns out that these coefficients do not satisfy the standard (unitarity) condition due to the following:

$$|\tilde{r}|^2 + |\tilde{t}|^2 = \frac{5}{9} < 1, \qquad \tilde{t} < 1 + \tilde{r} = \frac{4}{3}.$$

This informs us that the vertex conditions will not preserve the 'mass' of localized solutions $\mathcal{M} = \int_{X\in\mathbb{R}} |U|^2\,dX$.

3.2. NLS Soliton Scattering

We now consider the interaction of a bright soliton with the vertex. We integrate the NLS Equation (9) numerically using the fourth-order Runge-Kutta method. The Laplacian is discretized using a three-point central difference. Therefore, our scheme has the discretization error of at least order $\mathcal{O}(\Delta x^2, \Delta t^4)$, where Δx and Δt are the spatial mesh size and the time step, respectively. We used several combinations of Δx and Δt to make sure that variations of the computed results caused by the discretization are small enough. As the initial data, we take a soliton approaching the branch point from $X \ll 0$ in the form of $U(X,0)$ from (7), and without losing its generality, we set $A = 1$.

In Figure 2 we plot the dynamics of a soliton traveling towards the origin for two different initial velocities, namely $v = 0.3$ and 3, representing slow- and fast-moving solitons, respectively. In both cases, especially for the slow-incoming soliton, we see that as it approaches the branch point, it accelerates. This is usually a characteristic of an attractive potential. However, note that after the interaction, there is no trapped state, which on the other hand is a characteristic of a repulsive potential. It can be easily checked that the corresponding linear eigenvalue problem of (9) has no point spectrum, which confirms the absence of a trapped state. Thus, our branching point has both characteristics of an attractive as well as a repulsive potential at the same time.

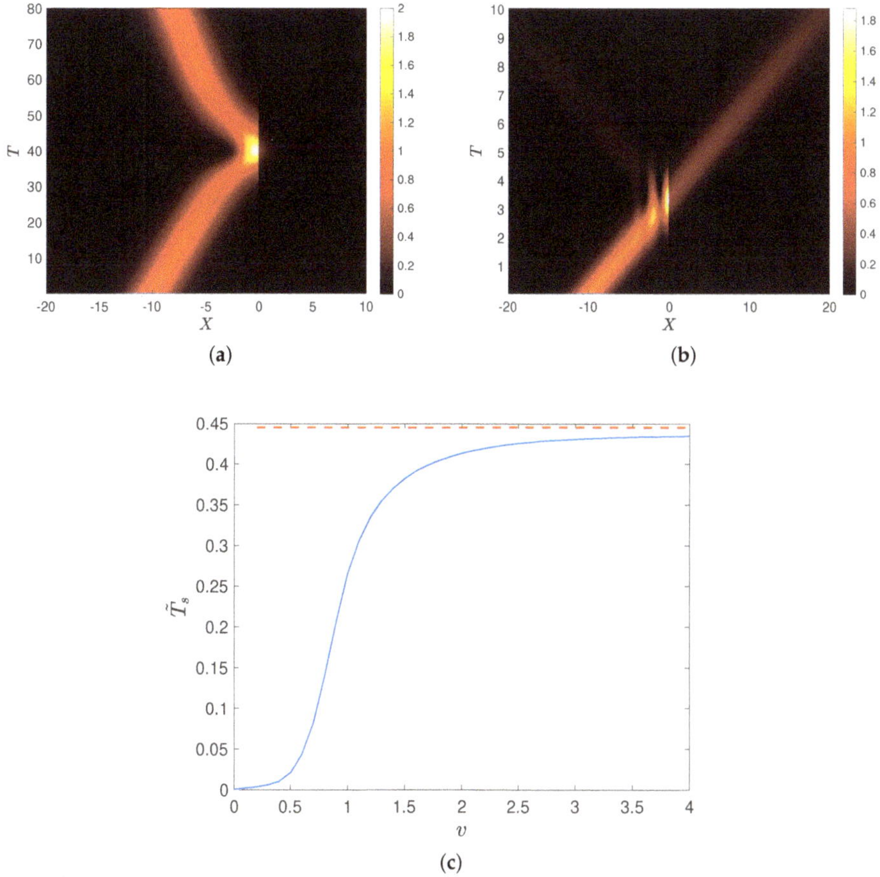

Figure 2. (**a**,**b**) Dynamics of a soliton moving towards the branch point $X = 0$ with the initial velocity $v = 0.3$ (**a**) and $v = 3$ (**b**). Shown are the top view of $|U(X,T)|^2$. (**c**) Numerically obtained transmission rate $\tilde{T}_s$ as a function of the incoming soliton velocity v. The horizontal dashed line is the theoretical approximation $\tilde{t}^2 = \frac{4}{9}$ from (11).

The main difference between Figure 2a,b is that for the slow-incoming soliton, most of the mass is reflected, while for the fast-moving one, most of it is transmitted. To quantify how much of the mass is being transmitted, we define the 'transmission rate' $\tilde{T}_s$ as

$$\tilde{T}_s = \lim_{T\to\infty} \frac{1}{4A} \int_{X>0} |U(X,T)|^2 \, dX. \tag{12}$$

Here, we normalize the rate with the initial L^2-norm $\lim_{T\to-\infty} ||U(T)||^2_{L^2} = 4A$, which (in the absence of the vertex conditions) normally would be constant in time. We plot the transmission rate $\tilde{T}_s$ in Figure 2c for different values of the initial velocity v. There is a steep transition in the interval of $v \approx 0.5$ and $v \approx 1.5$ where the soliton changes from being mostly reflected to mostly transmitted.

Recalling that for large v, the essential wave numbers of the Fourier transform of NLS solitons are concentrated around $k = v$, we can expect that the quantum transmission rate of a soliton with velocity v will approach a limiting value that is given by the absolute value square of the transmission coefficient of the linear plane wave $|\tilde{t}|^2$, see (11). We can see that this is indeed the case.

4. Scattering of sG Breathers

After studying the soliton scattering in the NLS setting, finally, we now consider the original problem, i.e., scattering in the sG equation context. The results of Section 3 should be comparable to the scattering of small-amplitude breathers in the sG equation.

In the infinite domain problem without any vertex condition at $x = 0$, the sG equation preserves the 'mass' $\mathcal{H} = \int_{x\in\mathbb{R}} H \, dx$, where the function H is the Hamiltonian given by

$$H(x,t) = \frac{1}{2}u_t^2 + \frac{1}{2}u_x^2 + (1 - \cos u). \tag{13}$$

We solved the sG Equation (8) using a similar numerical integration method previously implemented in solving the NLS equation in Section 3. As the initial data, we take the breather (4) with $x_0 \ll 0$ at $t = 0$.

First, we simulate the dynamics of small-amplitude breathers with $\theta = \cos^{-1} 0.99$. In Figure 3, we present our simulations of a breather traveling with two different velocities towards the branch point $x = 0$. It is interesting to see a close resemblance between panels (a, b) of Figure 3 and those of Figure 2. It is then instructive to compute the transmission rate of the breathers for different values of the incoming velocity. Defining

$$\tilde{T}_b = \lim_{T\to\infty} \frac{1}{\mathcal{H}_0} \int_{x>0} H(x,t) \, dx, \tag{14}$$

where $\mathcal{H}_0 = \lim_{t\to-\infty} ||H(t)||_{L^1}$, we plot in Figure 3c the transmission rate $\tilde{T}_b$, which again compared to Figure 2c shows the same qualitative profile. Moreover, we obtain numerically that the nonlinear transmission rate tends to the linear one $\tilde{t}^2$ (11) as the incoming breather velocity $v \to 1$.

Next, we consider large-amplitude breathers. In a similar setup, we simulated a slow- and a fast-incoming soliton. We present the typical dynamics of the two cases in Figure 4, where we take $\theta = \cos^{-1} 0.1$.

Far away from the branch point, a large-amplitude breather can be seen obviously as an oscillating pair of a kink and an anti-kink. Upon collision with the vertex, the slow-moving breather is trapped at the origin, while the fast-moving one dissociates into a trapped kink and an ejected anti-kink. Both cases show completely different dynamics from the small-amplitude breathers and the NLS solitons, see Figures 2 and 3. From the simulations, we note that the vertex acts as a repulsive potential for small-amplitude breathers, while it behaves as an attractive potential for large-amplitude ones. The difference can be explained as follows.

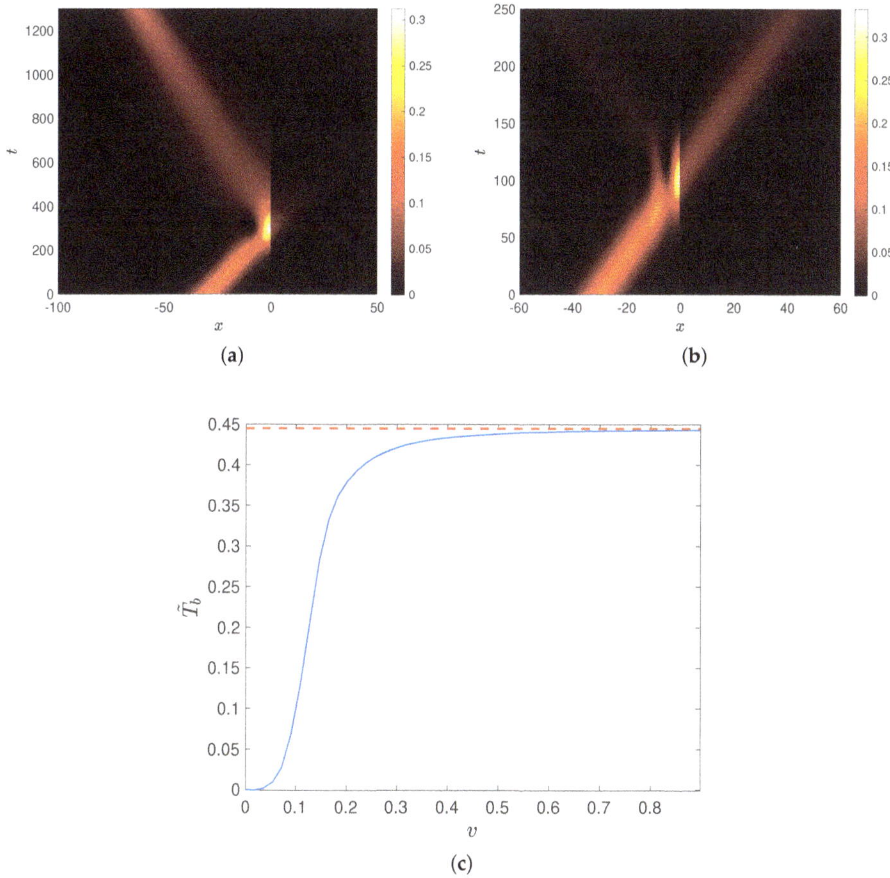

Figure 3. Similar to Figure 2, but for (small-amplitude) sG breathers with velocity $v = 0.1$ (**a**) and $v = 0.3$ (**b**). In panels (**a**,**b**), shown are the top view of $H(x,t)$. Panel (**c**) is the transmission rate $\tilde{T}_b$ as a function of the incoming breather velocity v. The horizontal dashed line is the theoretical approximation $\tilde{t}^2 = \frac{4}{9}$ from (11). Here, $\theta = \cos^{-1} 0.99$.

The branch point has been reported before to trap kinks, i.e., topological excitations [21,29,31,32]. They are given analytically by [29,31]

$$u = 4\tan^{-1}\left\{\exp[-x \pm \ln\sqrt{3}]\right\}, \tag{15}$$

with the '+' sign for the region $x < 0$ and the '−' sign for $x > 0$. They also have an oscillatory mode with frequency [29,31]

$$\omega = \sqrt{\frac{1+\sqrt{13}}{8}}. \tag{16}$$

When excited, the mode will vibrate, but eventually, it will fade away because of 'radiative' damping, i.e., a damping mechanism due to the excitation of higher harmonics with frequencies in the continuous spectrum. Even more, trapped kink oscillations decay asymptotically at a rate of $\mathcal{O}(t^{-1/2})$, see [34]. Here, even though our excitations are non-topological, large-amplitude breathers are close to an intertwining pair of kink and anti-kink. Trapping is therefore expected and in this case, an oscillatory mode must exist. Such oscillation can be seen quite well in Figure 4b where the trapped kink jiggles about $x = 0$.

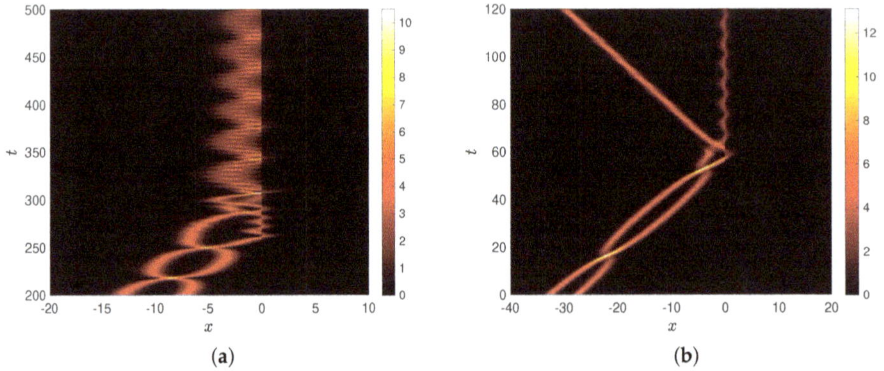

Figure 4. Similar to Figure 3a,b, but for a large-amplitude sG breather with velocity $v = 0.1$ (**a**) and $v = 0.5$ (**b**). Here, $\theta = \cos^{-1} 0.1$.

As for the oscillatory dynamics of the breather in Figure 4a, we can explain it as follows. The breather does not dissociate into a separate kink and anti-kink because its free energy is not enough to do so. As a result, it maintains its non-topological shape. In return, the vertex will tend to repel it. However, the non-topological excitation still has a relatively large amplitude, which on the other side looks like a kink and hence tends to be attracted by the vertex. Therefore, the branch point acts as a repellent and an attractor at the same time. This creates the oscillatory movement of the breather. Additionally, we also observe a fast oscillation. We strongly suspect that it is due to the trapping mode and hence its frequency is approximately given by (16).

5. Conclusions

We have analyzed for the first time the dynamics of breathers in a tricrystal Josephson junction. The physically relevant model consists of three semi-infinite Josephson junctions coupled at one end. For small-amplitude breathers, we have derived the corresponding NLS equation on star graphs from the governing sG equation. We have shown the resemblance of the dynamics of small-amplitude breathers with the NLS bright solitons. Large-amplitude breathers yield qualitatively different behaviors.

For future work, it will be interesting to rigorously study the collision of a fast-incoming NLS soliton with the vertex. This study will be along the lines of the analysis by Holmer et al. [19] about the scattering of fast-moving solitons by a delta interaction on the line. An extension of their work to the case of sG equations on graphs, which is not available yet, will also be particularly appealing. In the context of collective coordinate methods, it will be important to derive an effective Hamiltonian describing slow-moving soliton interactions with the vertex, along the idea of, e.g., [35–37]. The main challenge is how to incorporate the vertex condition, which is not explicitly embedded within the governing equation. On the numerical side, it will also be interesting to provide bounds to the discretization as well as the round-off errors of our numerical scheme.

Author Contributions: The contributions of the respective authors are as follows: H.S. designed and supervised the research and wrote the article; N.K. conducted analytical computations and wrote the article; Z.Z. performed numerical computations; T.N. and T.W. prepared the initial work and revised the article. All authors read and approved the final manuscript.

Funding: N.K. was funded by the National Research Foundation (NRF) of Korea through Grant No. NRF-2017-R1C1B5-017743 under the Basic Research Program in Science and Engineering.

Acknowledgments: The authors acknowledge the two referees for their comments that improved the paper and Andrew Harrison and Christopher Saker (University of Essex) for proofreading the manuscript.

Conflicts of Interest: The authors declare no conflict of interest.

Abbreviations

The following abbreviations are used in this manuscript:

NLS	nonlinear Schrödinger
sG	sine-Gordon

References

1. Pauling, L. *The Nature of the Chemical Bond and the Structure of Molecules and Crystals*; Cornell University Press: Ithaca, NY, USA, 1939.
2. Kottos, T.; Smilansky, U. Quantum chaos on graphs. *Phys. Rev. Lett.* **1997**, *79*, 4794–4797. [CrossRef]
3. Berkolaiko, G. An Elementary Introduction to Quantum Graphs. *Geom. Comput. Spectr. Theory* **2017**, *700*, 41–72.
4. Gnutzmann, S.; Smilansky, U. Quantum graphs: Applications to quantum chaos and universal spectral statistics. *Adv. Phys.* **2006**, *55*, 527–625. [CrossRef]
5. Berkolaiko, G.; Kuchment, P. *Introduction to Quantum Graphs, Volume 186 of Mathematical Surveys and Monographs*; American Mathematical Society: Providence, RI, USA, 2013.
6. Kuchment, P. Quantum graphs: An introduction and a brief survey. In *Analysis on Graphs and its Applications, Proceedings of Symposia in Pure Mathematics*; American Mathematical Society: Providence, RI, USA, 2008; pp. 291–314. Available online: https://arxiv.org/abs/0802.3442 (accessed on 8 January 2019)
7. Noja, D. Nonlinear Schrödinger equation on graphs: Recent results and open problems. *Philos. Trans. R. Soc. A Math. Phys. Eng. Sci.* **2014**, *372*, 20130002. [CrossRef] [PubMed]
8. Adami, R.; Cacciapuoti, C.; Finco, D.; Noja, D. On the structure of critical energy levels for the cubic focusing NLS on star graphs. *J. Phys. A. Math. Theor.* **2012**, *45*, 192001. [CrossRef]
9. Adami, R.; Cacciapuoti, C.; Finco, D.; Noja, D. Stationary states of NLS on star graphs. *EPL Europhys. Lett.* **2012**, *100*, 10003. [CrossRef]
10. Adami, R.; Cacciapuoti, C.; Finco, D.; Noja, D. Constrained energy minimization and orbital stability for the NLS equation on a star graph. *Ann. Inst. H. Poincaré C Anal. Non Linéare* **2014**, *31*, 1289–1310. [CrossRef]
11. Adami, R.; Cacciapuoti, C.; Finco, D.; Noja, D. Stable standing waves for a NLS on star graphs as local minimizers of the constrained energy. *J. Differ. Eqn.* **2015**, *260*, 7397–7415. [CrossRef]
12. Kairzhan, A. Orbital instability of standing waves for NLS equation on star graphs. *arXiv* **2017**, arXiv:1712.02773.
13. Li, Y.; Li, F.; Shi, J. Ground states of nonlinear Schrödinger equation on star metric graphs. *J. Math. Anal. Appl.* **2018**, *459*, 661–685. [CrossRef]
14. Noja, D.; Pelinovsky, D.E.; Shaikhova, G. Bifurcations and stability of standing waves in the nonlinear Schrödinger equation on the tadpole graph. *Nonlinearity* **2015**, *28*, 2343–2378. [CrossRef]
15. Marzuola, J.L.; Pelinovsky, D.E. Ground state on the dumbbell graph. *Appl. Math. Res. eXpress* **2016**, *2016*, 98–145. [CrossRef]
16. Goodman, R. NLS Bifurcations on the bowtie combinatorial graph and the dumbbell metric graph. *arXiv* **2017**, arxiv:1710.00030.
17. Noja, D.; Rolando, S.; Secchi, S. Standing waves for the NLS on the double-bridge graph and a rational–irrational dichotomy. *J. Differ. Eqn.* **2019**, *266*, 147–178. [CrossRef]
18. Adami, R.; Cacciapuoti, C.; Finco, D.; Noja, D. Fast solitons on star graphs. *Rev. Math. Phys.* **2011**, *23*, 409–451. [CrossRef]
19. Holmer, J.; Marzuola, J.; Zworski, M. Fast soliton scattering by delta impurities. *Commun. Math. Phys.* **2007**, *274*, 187–216. [CrossRef]
20. Sobirov, Z.; Matrasulov, D.; Sabirov, K.; Sawada, S.; Nakamura, K. Integrable nonlinear Schrödinger equation on simple networks: connection formula at vertices. *Phys. Rev. E* **2010**, *81*, 066602. [CrossRef] [PubMed]

21. Nakajima, K.; Onodera, Y.; Ogawa, Y. Logic design of Josephson network. *J. Appl. Phys.* **1976**, *47*, 1620–1627. [CrossRef]
22. Nakajima, K.; Onodera, Y. Logic design of Josephson network. II. *J. Appl. Phys.* **1978**, *49*, 2958–2963. [CrossRef]
23. Tsuei, C.C.; Kirtley, J.R.; Chi, C.C.; Yu-Jahnes, L.S.; Gupta, A.; Shaw, T.; Sun, J.Z.; Ketchen, M.B. Pairing symmetry and flux quantization in a tricrystal superconducting ring of $YBa_2Cu_3O_{7-\delta}$. *Phys. Rev. Lett.* **1994**, *73*, 593–596. [CrossRef] [PubMed]
24. Miller, J.H., Jr.; Ying, Q.Y.; Zou, Z.G.; Fan, N.Q.; Xu, J.H.; Davis, M.F.; Wolfe, J.C. Use of tricrystal junctions to probe the pairing state symmetry of $YBa_2Cu_3O_{7-\delta}$. *Phys. Rev. Lett.* **1995**, *74*, 2347–2350. [CrossRef] [PubMed]
25. Tsuei, C.C.; Kirtley, J.R. Phase-sensitive evidence for d-wave pairing symmetry in electron-doped cuprate superconductors. *Phys. Rev. Lett.* **2000**, *85*, 182–185. [CrossRef] [PubMed]
26. Tsuei, C.C.; Kirtley, J.R. Pairing symmetry in cuprate superconductors. *Rev. Mod. Phys.* **2000**, *72*, 969–1016. [CrossRef]
27. Tomaschko, J.; Scharinger, S.; Leca, V.; Nagel, J.; Kemmler, M.; Selistrovski, T.; Koelle, D.; Kleiner, R. Phase-sensitive evidence for $d_{x^2-y^2}$-pairing symmetry in the parent-structure high-Tc cuprate superconductor $Sr_{1-x}La_xCuO_2$. *Phys. Rev. B* **2012**, *86*, 094509. [CrossRef]
28. Kogan, V.G.; Clem, J.R.; Kirtley, J.R. Josephson vortices at tricrystal boundaries. *Phys. Rev. B* **2000**, *61*, 9122–9129. [CrossRef]
29. Susanto, H.; van Gils, S.A. Existence and stability analysis of solitary waves in a tricrystal junction. *Phys. Lett. A* **2005**, *338*, 239–246. [CrossRef]
30. Susanto, H.; van Gils, S.A.; Doelman, A.; Derks, G. Analysis on the stability of Josephson vortices at tricrystal boundaries: A $3\phi_0/2$-flux case. *Phys. Rev. B* **2004**, *69*, 212503, 1–4. [CrossRef]
31. Grunnet-Jepsen, A.; Fahrendorf, F.N.; Hattel, S.A.; Grønbech-Jensen, N.; Samuelsen, M.R. Fluxons in three long coupled Josephson junctions. *Phys. Lett. A* **1993**, *175*, 116–120. [CrossRef]
32. Hattel, S.A.; Grunnet-Jepsen, A.; Samuelsen, M.R. Dynamics of three coupled long Josephson junctions. *Phys. Lett. A* **1996**, *221*, 115–123. [CrossRef]
33. Krämer, P. The Method of Multiple Scales for nonlinear Klein-Gordon and Schrödinger Equations. Diploma Thesis, Karlsruhe Institute of Technology, Karlsruhe, Germany, 2013.
34. Ali, A.; Susanto, H.; Wattis, J.A. Breathing modes of long Josephson junctions with phase-shifts. *SIAM J. Appl. Math.* **2011**, *71*, 242–269. [CrossRef]
35. Forinash, K.; Peyrard, M.; Malomed, B. Interaction of discrete breathers with impurity modes. *Phys. Rev. E* **1994**, *49*, 3400–3411. [CrossRef]
36. Goodman, R.H.; Holmes, P.J.; Weinstein, M.I. Strong NLS soliton–defect interactions. *Phys. D* **2004**, *192*, 215–248. [CrossRef]
37. Holmer, J.; Zworski, M. Slow soliton interaction with delta impurities. *J. Mod. Dyn.* **2007**, *1*, 689–718.

symmetry

Article

Coupling Conditions for Water Waves at Forks

Jean–Guy Caputo [1,*], Denys Dutykh [2] and Bernard Gleyse [1]

1 Laboratoire de Mathématiques, INSA Rouen Normandie, 76801 Saint–Etienne du Rouvray, France; gleyse@insa-rouen.fr
2 University Grenoble Alpes, University Savoie Mont Blanc, CNRS, LAMA, 73000 Chambéry, France; Denys.Dutykh@univ-savoie.fr
* Correspondence: caputo@insa-rouen.fr

Received: 15 January 2019; Accepted: 19 March 2019; Published: 24 March 2019

Abstract: We considered the propagation of nonlinear shallow water waves in a narrow channel presenting a fork. We aimed at computing the coupling conditions for a 1D effective model, using 2D simulations and an analysis based on the conservation laws. For small amplitudes, this analysis justifies the well-known Stoker interface conditions, so that the coupling does not depend on the angle of the fork. We also find this in the numerical solution. Large amplitude solutions in a symmetric fork also tend to follow Stoker's relations, due to the symmetry constraint. For non symmetric forks, 2D effects dominate so that it is necessary to understand the flow inside the fork. However, even then, conservation laws give some insight in the dynamics.

Keywords: networks; nonlinear shallow water equations; nonlinear wave equations

1. Introduction

The propagation of nonlinear waves in a network is an important topic. As an example, consider a hydrological network which is prone to floods. Understanding the global dynamics of the network can help identify its most vulnerable sections and take the appropriate measures. Real networks are formed by long 2D or 3D channels of a small cross-section. To study the propagation of waves in such systems, a first step is to consider a simple fork as a model of elementary junctions. The final goal is to reduce the model to 1D channels connected by appropriate interface conditions. The study of such 1D systems is now well advanced, in particular for systems of conservation laws, see the review [1].

The type of PDE model describing the quantity propagating on the network is very important to derive the coupling conditions. Recently for the sine-Gordon nonlinear wave equation, we [2] introduced a homothetic reduction [3] where we averaged the operator over the fork region and consistently took the limit when the width tended to zero. Assuming continuity of the field, we obtained Kirchhoff's law for the gradients. Comparing the 2D solution with the one for the reduced 1D equations gives excellent agreement. In this situation, the angle of the fork does not play a role. When considering networks of rivers, many authors, for example Stoker [4] and Jacovkis [5] assumed continuity of the water height and continuity of the flux so that again, the angle of the fork did not come in. In the close context of gas dynamics, Holden and Risebro [6] studied shocks in a pipe with an elbow. They showed that the Riemann problem had a unique solution when the angle was smaller than π. The angle is also important for classical hydrodynamics; in a fork, it sets the forces experienced by the pipes [7]. In fact, for large amplitude shallow water waves our numerical calculations show that the energy entering a branch can vary from 20% to 50% depending on the symmetry of the fork. These studies point out the importance of the angle.

A few authors addressed the problem of the angle of a junction. Schmidt [8] studied the 2D connection between 1D channels; he made no assumption on the size of the connecting domain. The flow in the junction was assumed linear so that the author used a variational method that gave the solution as a superposition of fields. The final result was a system of ordinary differential equations for the values at the ends of the branches coupled to the shallow water PDEs. Despite its formal beauty, it remains difficult to handle and does not give a simple picture. Shi et al. [9] studied experimentally and numerically the propagation of long waves in wide and narrow channels. They used the Boussinesq dispersive shallow water equations for narrow channels. They observed no angle dependence and a strong transmission. For the same equations, Nachbin and Simoes [10] obtained interface conditions containing implicitly the angles of the fork. These gave an excellent matching between the average of the 2D solution and the solution of the 1D effective model for angles smaller than $\pi/3$.

In this article, we consider the nonlinear shallow water equations. The system is very general because it only involves conservation laws. Also it is simple enough. We revisit the problem of shallow water propagation in 2D forks using our homothetic reduction procedure to obtain approximate conservation laws and compare them with the numerical solutions. We compute approximate conservation for the mass, momenta and energy laws for a general fork geometry. In the small amplitude limit we recover Stoker's conditions, i.e., continuity of surface elevation and mass conservation (Kirchoff law). To our knowledge, this is a first formal justification of Stoker's interface conditions. This angle independent reduction holds also for a general class of scalar nonlinear wave equations, for example the 2D sine-Gordon equation or the 2D reaction-diffusion equation; it confirms the results of [2]. We computed the 2D numerical solution for a simple T-fork geometry for small and large amplitudes. The wave was also launched in two different branches to see the effect of symmetry. We show that Stoker's conditions hold for the symmetric case for small and large amplitudes. For the non-symmetric case, they hold for small amplitudes. When the amplitude is large, 2D effects dominate the fork region. Nevertheless the approximate conservation laws give an insight into the flow.

The article is organized as follows. Section 2 presents the fork geometry and shows the straightforward reduction for a general class of nonlinear wave equations. In Section 3 we recall the shallow water equations and their conserved quantities. Section 4 gives the integrals of these equations on the fork showing that the mass and energy laws do not involve the angles while the momenta laws do. Section 5 shows the 2D numerical solutions for symmetric and non symmetric configurations for small and large waves. There, we compare the numerical results with the conservation laws established in Section 4. We discuss these results and conclude in Section 6.

2. General Scalar Nonlinear Wave Equations

Before considering the nonlinear shallow water equations, we analyze the simpler case of a class of scalar 2D nonlinear wave equations. This large class includes hyperbolic wave equations like the sine-Gordon equation as well as reaction diffusion equations like the Fisher equation, to name a few. We consider equations of the form

$$\alpha u_{tt} + \beta u_t - \Delta u = N(u), \tag{1}$$

where $u(x, y, t)$ is a scalar, Δ is the usual 2D Laplacian and where $N(u)$ is a nonlinearity not containing derivatives. The boundary condition on the lateral domain is of Neumann type

$$\partial_n u = \nabla u \cdot \mathbf{n} = 0. \tag{2}$$

Consider the fork domain shown in Figure 1. Far from the fork region, the solution can be assumed to be 1D so that we do not loose much information by approximating the 2D dynamics with a 1D equation. Inside the fork domain, a strong coupling occurs between the branches. To see this, we proceed as in [2] and integrate the operators on the fork region. Then we examine the behavior of the different terms as w, the width of the branches, goes to zero. We assume that domains that we consider behave in a regular way as we shrink w homothetically to zero, [3].

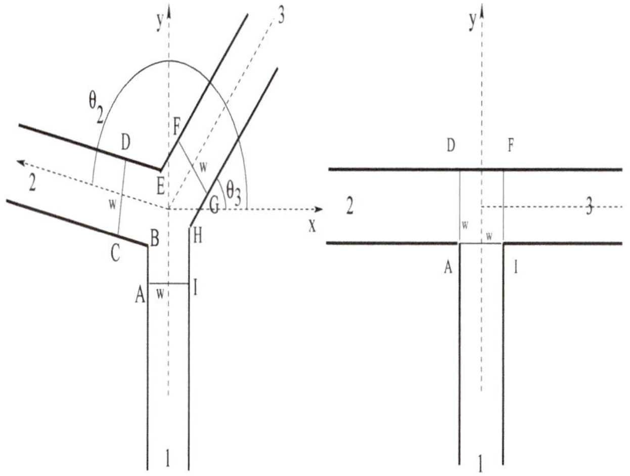

Figure 1. A fork geometry with arbitrary angles (**left**) and with right angles (**right**).

Consider the asymmetric Y-branch shown in the left panel of Figure 1. A first assumption is the continuity of u which is obvious for the 2D operator. The other condition comes from the integration of the operator (1) on the fork domain $\mathcal{F} = IABCDEFGHI$. We get

$$\int [\alpha u_{tt} + \beta u_t - N(u)]\, dxdy - \int_{\partial \mathcal{F}} (\nabla u) \cdot \mathbf{n}\, ds = 0. \tag{3}$$

The first integral is of order $O(w^2)$. On the exterior boundaries, $(\nabla u) \cdot \mathbf{n} = 0$ so the line integral reduces to

$$\int_{IA} \cdots + \int_{CD} \cdots + \int_{FG} \cdots,$$

which are $O(w)$. We then obtain for $w \to 0$

$$-\partial_s u_1 + \partial_s u_2 + \partial_s u_3 = 0, \tag{4}$$

where u_i, $i = 1,2,3$ are respectively the values of the field at the end of branch 1 (IA) and at the beginning of branches 2 (FG) and 3 (CD). Relation (4) is Kirchhoff's law [2]. When the widths of the branches are not equal, this Kirchoff relation becomes

$$-w_1 \partial_s u_1 + w_2 \partial_s u_2 + w_3 \partial_s u_3 = 0. \tag{5}$$

Remark that in the result (4) the angle of the fork plays no role. The reduction leading from the flux equation to (5) is an asymptotic result that holds for $w \to 0$. It is then natural to approximate the 2D Equation (1) by a 1D equation in each branch together with the conditions of continuity and Kirchoff (4) at the junctions.

The result we obtain can be connected to a property of the Laplace operator with Neumann boundary conditions on a so-called "fat" graph [11]. Consider a graph where each edge has a transverse size w, assume Neumann boundary conditions on the transverse edge. Then the spectrum of the Laplacian converges to the one of the 1D Laplacian as $w \to 0$. This is true for compact and non compact graphs. See the article by Exner and Post [11] and the book by Post [12] for the details of the proof.

The validity of the reduction was confirmed numerically for the 2D sine-Gordon equation, (1) with $\alpha = 1$, $\beta = 0$ and $N(u) = -\sin(u)$ in [2]. There we compared the 2D solutions to the ones of the 1D sine-Gordon equation in each branch, coupled by the interface conditions. For completeness, we recall the case of a sine-Gordon kink propagating in forks with angles 45 and 90 degrees. The kink is an exact solution in 1D, it is

$$u(x,t) = 4\arctan\left[\exp(\frac{x-vt}{\sqrt{1-v^2}})\right], \tag{6}$$

where the velocity $0 \le v < 1$ is a free parameter. To compare the 2D and 1D solutions, we plot the energies in each branch

$$E_2^i = \int_{\Omega_i} \left[\frac{1}{2}u_t^2 + \frac{1}{2}|\nabla u|^2 + (1-\cos u)\right] dxdy, \tag{7}$$

and

$$E_1^i = \sum_{i=1,2,3} \int_{\Omega_i} \left[\frac{1}{2}u_t^2 + \frac{1}{2}|u_x|^2 + (1-\cos u)\right] dx, \tag{8}$$

where Ω_i is branch i, abusively named the same in 1D and 2D. The kink is started in branch 1 with an initial velocity $v = 0.75$, this gives a typical wavelength $\lambda \approx 4/\sqrt{1-v^2} = 2.7$. The width of the branches is $w = 0.7 << \lambda$. Figure 2 shows the time evolution of the energies E_2^i for forks with angles 45 and 90 degrees and E_1^i, where $i = 1,2$ corresponds to the branches. Initially the kink is in branch 1 so that $E_2^2 = E_2^3 = 0$. As the kink crosses into branches 2 and 3, E_2^1 becomes very small. Note the excellent agreement between the two expressions E_2^i and the expression E_1^i. This confirms that the angle of the fork plays no role for such a system.

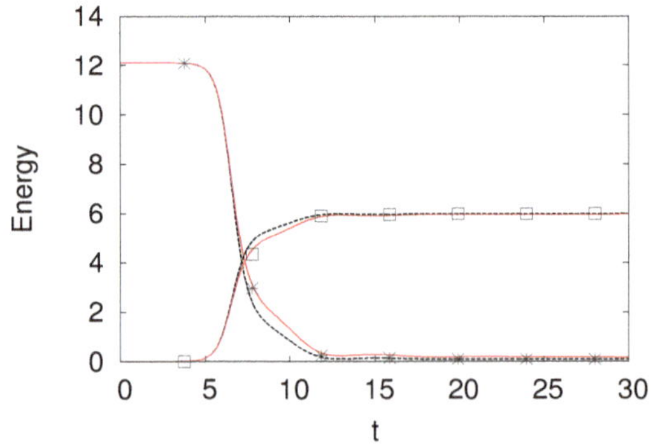

Figure 2. Time evolution of the energies E_2^i for the kink motion in branches $i = 1,2$ for the T-junction (90 degrees) in full line (red online), for the Y-junction (45 degrees) in dashed line. The energy E_1^i for the 1D effective model is plotted with points.

The dynamics of kinks for the sine-Gordon equation is controlled by the energy: if the initial energy is enough, a kink in branch 1 gives rise to two kinks in branches 2 and 3. This gives a very simple picture. Other solutions like the breather have much more complicated dynamics, we refer the reader to [2] for more details.

The dynamics of such waves can then be studied for general networks as we have done in [13].

3. The Nonlinear Shallow Water Equations

The shallow water equations in a 2D domain written in terms of the fluid velocity $\mathbf{u}(x,t)$

$$\mathbf{u} = (u,v)^T$$

and the water height $h(x,t)$ read [4]

$$h_t + \nabla \cdot (h\mathbf{u}) = 0, \tag{9}$$

$$(hu)_t + \nabla \cdot \begin{pmatrix} hu^2 + \frac{gh^2}{2} \\ huv \end{pmatrix} = 0, \tag{10}$$

$$(hv)_t + \nabla \cdot \begin{pmatrix} huv \\ hv^2 + \frac{gh^2}{2} \end{pmatrix} = 0, \tag{11}$$

where g is the gravitational acceleration. The wall boundary condition is

$$\mathbf{u} \cdot \mathbf{n} = 0. \tag{12}$$

We assume an even bottom of the channels $h = h_0$.

3.1. Conserved Quantities

We first recall the conserved quantities. Integrating Equations (9)–(11) over a 2D closed domain Ω and using the boundary condition (12) we get

$$\partial_t \int_\Omega h\, dxdy = 0, \tag{13}$$

$$\partial_t \int_\Omega hu\, dxdy + \oint_{\partial\Omega} \frac{gh^2}{2} n_x\, ds = 0, \tag{14}$$

$$\partial_t \int_\Omega hv\, dxdy + \oint_{\partial\Omega} \frac{gh^2}{2} n_y\, ds = 0. \tag{15}$$

A localized wave will have as first conserved quantity the integral of the water elevation

$$M = \int_\Omega h\, dxdy.$$

The total x and y momenta

$$P_x = \int_\Omega hu\, dxdy, \quad P_y = \int_\Omega hv\, dxdy$$

will not be conserved in the fork geometries.

A flux relation that can be deduced from the conservation laws (9)–(11) is the total energy flux

$$e_t + \nabla \cdot \left[\mathbf{u}(e + \frac{gh^2}{2}) \right] = 0. \tag{16}$$

where the total energy density is

$$e = \frac{1}{2} \left[gh^2 + (u^2 + v^2)h \right]. \tag{17}$$

Integrating the energy flux relation over a volume Ω we obtain that a localized wave in Ω will have constant energy

$$\frac{dE}{dt} = \frac{d}{dt} \int_\Omega e\, dxdy = 0.$$

3.2. Small Amplitude Limit

It is well known that in the linear limit, Equations (9)–(11) reduce to the linear wave equation for the water height h. To see this, consider the steady state $h = h_0,\ u = v = 0$, then the linearized system is

$$h_t + h_0 \nabla \mathbf{u} = 0, \tag{18}$$

$$h_0 \mathbf{u}_t + g \nabla h = 0, \tag{19}$$

Taking the time derivative of the first equation and plugging in the second equation, we get the wave equation

$$h_{tt} - g h_0 (h_{xx} + h_{yy}) = 0. \tag{20}$$

The boundary conditions reduce to $\nabla h \cdot \mathbf{n} = 0$ as can be seen by projecting (19) on $\mathbf{n}$. This equation is in the class (1).

4. Reduction of the Shallow Water Equations

The shallow water equations cannot be reduced so simply as the nonlinear scalar wave equation. In fact, it is not clear what are the right interface conditions that should be implemented for a 1D effective model. Stoker, in his well-known book [4] introduces the following interface conditions for the water elevations h_1, h_2, h_3 and branch-oriented velocities $u_1^{\|}, u_2^{\|}, u_3^{\|}$

$$h_1 = h_2 = h_3, \tag{21}$$

$$-h_1 u_1^{\|} + h_2 u_2^{\|} + h_3 u_3^{\|} = 0, \tag{22}$$

and uses them to analyze the junction of the Mississippi and the Missouri rivers. These conditions were not justified by a formal argument. Note also that they do not depend on the angle of the junction.

Below, we will see that these conditions arise naturally in the limit of small amplitude for the shallow water equations. For general amplitudes, it is not clear that these apply. To analyze the problem, we proceed as in [2], integrate the governing equations on the bifurcation region and consider the limit of vanishing transverse width w.

4.1. Mass Flux

Integrating the Equation (9) over the closed region $\mathcal{F} \equiv ABCDEFGHIA$ yields

$$\int_{\mathcal{F}} h_t \, dxdy + \oint_{\partial \mathcal{F}} h \, \mathbf{u} \cdot \mathbf{n} \, ds = 0.$$

Because of the boundary condition $\mathbf{u} \cdot \mathbf{n} = 0$ on ABC, DEF and GHI the expression above reduces to

$$\int_{\mathcal{F}} h_t \, dxdy + \int_{AI} h \, \mathbf{u} \cdot \mathbf{n} \, ds + \int_{CD} h \, \mathbf{u} \cdot \mathbf{n} \, ds + \int_{FG} h \, \mathbf{u} \cdot \mathbf{n} \, ds = 0.$$

The first integral is $O(w^2)$ while the three other integrals are $O(w)$. Dividing the equation by w and taking the limit $w \to 0$ we get from these three terms

$$-h_1 u_1^{\|} + h_2 u_2^{\|} + h_3 u_3^{\|} = 0, \tag{23}$$

where we have introduced the local branch-oriented velocities $u^{\|}, u^{\perp}$ such that

$$\begin{pmatrix} u^{\|} \\ u^{\perp} \end{pmatrix} = \begin{pmatrix} \cos\theta & \sin\theta \\ -\sin\theta & \cos\theta \end{pmatrix} \begin{pmatrix} u \\ v \end{pmatrix} \tag{24}$$

and where the indices 1,2 and 3 refer to the branches. Of course, when the transverse widths w_1, w_2, w_3 are different, with the condition that the ratios $w_2/w_1, w_3/w_1$ remain finite, the relation (23) becomes

$$-w_1 h_1 u_1^{\parallel} + w_2 h_2 u_2^{\parallel} + w_3 h_3 u_3^{\parallel} = 0.$$

4.2. Energy Flux

The energy flux (16) can be consistently reduced to a 1D relation. As for the mass relation, we integrate Equation (17) over the domain $\mathcal{F}$ = ABCDEFGHIA to obtain

$$\int_{\mathcal{F}} e_t \, dxdy + \oint_{\partial\mathcal{F}} (e + \frac{gh^2}{2}) \, \mathbf{u} \cdot \mathbf{n} \, ds = 0.$$

Because of the boundary condition $\mathbf{u} \cdot \mathbf{n} = 0$ on ABE, the expression above reduces to

$$\int_{\mathcal{F}} e_t \, dxdy + \int_{AI} (e + \frac{gh^2}{2}) \, \mathbf{u} \cdot \mathbf{n} \, ds + \int_{CD} (e + \frac{gh^2}{2}) \, \mathbf{u} \cdot \mathbf{n} \, ds + \int_{FG} (e + \frac{gh^2}{2}) \, \mathbf{u} \cdot \mathbf{n} \, ds = 0.$$

The first integral is $O(w^2)$ while the three other integrals are $O(w)$. Dividing the equation by w and taking the limit $w \to 0$ we get from these three terms

$$-(e_1 + \frac{gh_1^2}{2}) u_1^{\parallel} + (e_2 + \frac{gh_2^2}{2}) u_2^{\parallel} + (e_3 + \frac{gh_3^2}{2}) u_3^{\parallel} = 0. \tag{25}$$

To conclude, Equation (9) gives in the 1D limit, the balance of mass (23). The same happens for the energy flux (16) which yields (25). The natural matching conditions for 1D shallow water equations on a network are then

$$-h_1 u_1^{\parallel} + h_2 u_2^{\parallel} + h_3 u_3^{\parallel} = 0, \tag{26}$$

$$-u_1^{\parallel} (g h_1^2 + h_1 \frac{{u_1^{\parallel}}^2}{2}) + u_2^{\parallel} (g h_2^2 + h_2 \frac{{u_2^{\parallel}}^2}{2}) + u_3^{\parallel} (g h_3^2 + h_3 \frac{{u_3^{\parallel}}^2}{2}) = 0. \tag{27}$$

For the mass and the energy balance laws, we have a similar situation to the one of the nonlinear scalar wave equation, the angles of the fork do not play any role. In the small amplitude limit, the speeds u_1, u_2, u_3 are small and the squares can be neglected in the energy relation. Then, we recover the Stoker interface conditions (21).

4.3. Momentum Flux for a General Fork

Contrary to the mass and the energy, the momentum Equations (10) and (11) cannot be consistently reduced to a 1D condition involving $h, u^{\parallel}$ at each end of $\mathcal{F}$.

To see this, integrate the horizontal momentum Equation (10) over the domain $\mathcal{F}$ and get

$$\int_{\mathcal{F}} (hu)_t \, dxdy + \oint_{\partial\mathcal{F}} \begin{pmatrix} hu^2 + \frac{gh^2}{2} \\ huv \end{pmatrix} \cdot \mathbf{n} \, ds = 0,$$

where the first integral is a surface integral and the second one a line integral. In the integrand of the latter, we have

$$\begin{pmatrix} hu^2 \\ huv \end{pmatrix} \cdot \mathbf{n} = hu \begin{pmatrix} u \\ v \end{pmatrix} \cdot \mathbf{n} = 0$$

on the exterior boundaries of $\partial\mathcal{F}$ because of the boundary condition (2). Then, only the potential term $\frac{gh^2}{2}$ will contribute to these terms.

The $O(w)$ terms (line integrals) reduce to

$$\begin{aligned}&-\tfrac{g}{2}(|AB|h_{AB}^2-|HI|h_{HI}^2)-\sin\theta_2\tfrac{g}{2}(|BC|h_{BC}^2-|DE|h_{DE}^2)-\sin\theta_3\tfrac{g}{2}(|EF|h_{EF}^2-|HG|h_{HG}^2)\\&-wh_1u_1v_1+w\left[(h_2u_2^2+g\tfrac{h_2^2}{2})\cos\theta_2+h_2u_2v_2\sin\theta_2\right]+w\left[(h_3u_3^2+g\tfrac{h_3^2}{2})\cos\theta_3+h_3u_3v_3\sin\theta_3\right]=0.\end{aligned} \tag{28}$$

Using the branch oriented velocities (24) we get the approximate law

$$\begin{aligned}&-\tfrac{g}{2}(|AB|h_{AB}^2-|HI|h_{HI}^2)-\sin\theta_2\tfrac{g}{2}(|BC|h_{BC}^2-|DE|h_{DE}^2)-\sin\theta_3\tfrac{g}{2}(|EF|h_{EF}^2-|GH|h_{GH}^2)\\&-wh_1u_1v_1+w\,\cos\theta_2\left[h_2{u_2^{\parallel}}^2+g\tfrac{{h_2}^2}{2}\right]+w\,\cos\theta_3\left[h_3{u_3^{\parallel}}^2+g\tfrac{{h_3}^2}{2}\right]=0,\end{aligned} \tag{29}$$

where we neglected the velocity components $u^{\perp}$.

Similarly for the vertical momentum equation we obtain

$$\begin{aligned}&\frac{g}{2}\cos\theta_2(|BC|\,h_{BC}^2-|DE|\,h_{DE}^2)+\frac{g}{2}\cos\theta_3(|EF|\,h_{EF}^2-|GH|\,h_{GH}^2)-w\,\left[h_1v_1^2+g\frac{{h_1}^2}{2}\right]\\&+w\left[(h_2u_2^2+g\frac{h_2^2}{2})\sin\theta_2+h_2u_2v_2\cos\theta_2\right]+w\left[(h_3u_3^2+g\frac{h_3^2}{2})\sin\theta_3+h_3u_3v_3\cos\theta_3\right]=0.\end{aligned} \tag{30}$$

Using the branch velocities and neglecting the transverse components we get

$$\begin{aligned}&\frac{g}{2}\cos\theta_2(|BC|\,h_{BC}^2-|DE|\,h_{DE}^2)+\frac{g}{2}\cos\theta_3(|EF|\,h_{EF}^2-|GH|\,h_{GH}^2)\\&-w\left[h_1v_1^2+g\frac{{h_1}^2}{2}\right]+w\sin\theta_2\left[h_2{u_2^{\parallel}}^2+g\frac{h_2^2}{2}\right]+w\sin\theta_3\left[h_3{u_3^{\parallel}}^2+g\frac{h_3^2}{2}\right]=0.\end{aligned} \tag{31}$$

4.4. Momentum Flux for the T-Fork

Consider now the T-geometry shown in the right panel of Figure 1. The calculations are simpler so that we used this geometry to validate the approach numerically. The general fork domain $\mathcal{F}$ can be reduced to the square $ADFIA$ by taking $\theta_2=\pi,\theta_3=0$ and $B\to C\to A,\;\; G\to H\to I$. Then the Equations (28) and (30) reduce to

$$-h_1u_1v_1-(h_2u_2^2+g\frac{h_2^2}{2})+h_3u_3^2+g\frac{h_3^2}{2}=0, \tag{32}$$

$$-(h_1v_1^2+g\frac{h_1^2}{2})-h_2u_2v_2+g\frac{h_{23}^2}{2}+h_3u_3v_3=0, \tag{33}$$

where the term h_{23} is

$$h_{23}^2\equiv\frac{1}{w}\int_{DF}h^2\,ds. \tag{34}$$

We will see that it can be obtained by interpolation of h_2 and h_3.

4.5. Effective 1D Model for the T-Fork

The pseudo-conservation laws (26), (27), (29) and (31) established in the previous section in the limit $w\to 0$ provide a formal connection between $h,u^{\parallel}$ in branches 1,2 and 3. In principle, they enable to approximate the 2D problem (9)–(11) by three 1D shallow water equations

$$H_t^i+(H^iU^i)_x=0, \tag{35}$$

$$(H^iU_i)_t+(H^i{U^i}^2+\frac{gH_i^2}{2})_x=0, \tag{36}$$

where $i = 1, 2, 3$ correspond to the different branches. These 1D shallow water equations can be solved using a standard finite difference scheme, see for example [14]. The discretization is shown in Figure 3 where the first nodes in each branch have values $H = h_i, U = u_i$. The coupling equations between these three nodes given by (26), (27) and (29) would be solved using a Newton iteration.

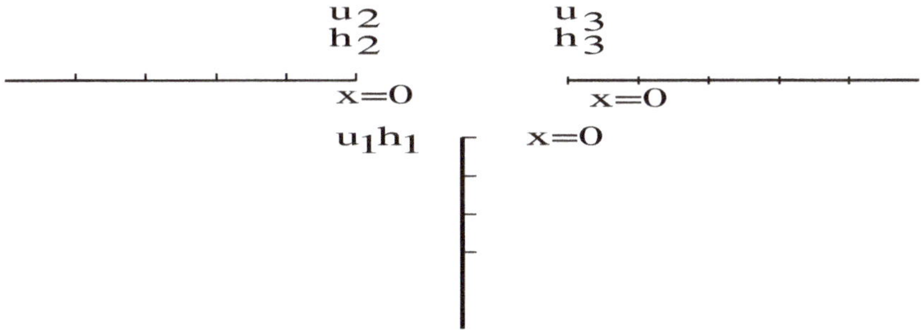

Figure 3. Space discretization for the 1D approximation.

5. Numerical Solutions of the 2D Shallow Water Equations

The approximation described in the previous section holds if the error remains small. We now evaluate this error by solving numerically the 2D problem (9)–(11), compute $h, u^{\|}$ and see how these values agree with the pseudo-conservation laws (26), (27), (29) and (31). We chose the T geometry shown in the right panel of Figure 1 for simplicity and considered symmetric and non symmetric initial conditions. We also increased the wave amplitude to estimate the effect of the non linearity.

The Equations (9)–(11) were discretized using as space unit the depth d. The time unit was $\sqrt{\frac{d}{g}}$. The variables and fields was rescaled as

$$x' = \frac{x}{d}, \quad t' = t\sqrt{\frac{g}{d}}, \quad h' = \frac{x}{d}, \quad u' = \frac{u}{\sqrt{gd}}. \tag{37}$$

This amounts to taking $d = 1, \ g = 1$ in (9)–(11).

We solved the nonlinear shallow water equations using a first order finite volume scheme on an unstructured triangular mesh produced with the Gmsh meshing software (see details in [15]). We used the width $w = 0.125$ and the typical size of the triangles is 0.02. The time advance used a variable order Adams–Bashforth–Moulton multistep solver (implemented in Matlab under ode113 subroutine [16]). The relative and absolute tolerances were set to 10^{-5}.

The initial condition is taken as a travelling solitary wave of velocity c. This is an exact solution for the mass conservation law. We used a solitary wave inspired by the Serre theory [17], (see [18] for the modern variational derivation)

$$h(x,y,t=0) = d + \eta(y), \tag{38}$$

$$v(x,y,t=0) = c\frac{\eta(y)}{d+\eta(y)}, \tag{39}$$

$$\eta(y) = a \,\mathrm{sech}^2(\frac{1}{2}k(y-y_0)), \tag{40}$$

where the speed is

$$c = \sqrt{g(d+a)}.$$

The other parameters were

$$g = 1, \ k = 1, \ d = 1, \ a = 1, \ x_0 = y_0 = 2.5.$$

The wave was chosen so that its extension $2/k = 2$ is much larger than the width $w = 0.125$. Below we discuss the effect of the width.

The four pseudo-conservation laws for the mass, momenta and energy (26), (27) and (29) on the fork domain $ADFIA$ are

$$\delta m \equiv -h_1 v_1 - h_2 u_2 + h_3 u_3 = 0, \tag{41}$$

$$\delta p_x \equiv -h_1 u_1 v_1 - (h_2 u_2^2 + g\frac{h_2^2}{2}) + h_3 u_3^2 + g\frac{h_3^2}{2} = 0. \tag{42}$$

$$\delta p_y \equiv -(h_1 v_1^2 + g\frac{h_1^2}{2}) - h_2 u_2 v_2 + g\frac{h_{23}^2}{2} + h_3 u_3 v_3 = 0, \tag{43}$$

$$\delta e \equiv -v_1(g h_1^2 + h_1\frac{v_1^2}{2}) - u_2(g h_2^2 + h_2\frac{u_2^2}{2}) + u_3(g h_3^2 + h_3\frac{u_3^2}{2}) = 0, \tag{44}$$

where we introduced the residuals δm, δp_x, δp_y and δe.

We considered a symmetric situation where the wave is incident from branch 1 and a non symmetric situation where the wave was send into the fork from branch 3. In both cases, the number of unknowns was the same; see Table 1.

Table 1. The two different dynamic problems for the T-branch.

Type	Known	Unknown
wave in branch 1	h_1, v_1	h_2, u_2, h_3, u_3
wave in branch 3	h_3, u_3	h_1, v_1, h_2, u_2

The wave mass and wave energy in each branch have been calculated. They are defined as

$$M_w = \int_\Omega (h - d)\ dxdy,$$

$$E_w = \int_\Omega \frac{1}{2}\left[g(h-d)^2 + (u^2 + v^2)h\right]\ dxdy.$$

Energy will propagate very differently in problems 1 and 2. In the next sections we examine in detail the two types of problems and use the conservation laws to establish jump conditions for the 1D effective model.

To verify the approximation given by the relations (41)–(44), we also computed the time evolution of the quantities $h_1, h_2, h_3, v_1, u_2, u_3$ from the 2D direct numerical simulations. We used a scattered linear interpolation to estimate these physical variables along the four different segments of the fork region from the unstructured triangular mesh data.

5.1. Wave Incident into Branch 1

5.1.1. Small Amplitude Waves $a/d = 0.1$

The time evolution of the wave mass and energy is presented in Figure 4. Consider the wave mass, at $t = 0$: $M_1^0 = 57\ 10^{-3}, M_2^0 = M_3^0 = 0$. After the wave has passed, at $t = 6.5, M_1 = 0, M_2 = M_3 = 26$. We have $2 \times 26 = 52$ which shows the conservation of mass. Notice the depression in the mass in branch 1 after the wave passes. Almost all energy is transferred to branches 2 and 3.

Here our balance laws hold well for both the mass and the energy, see Figure 5.

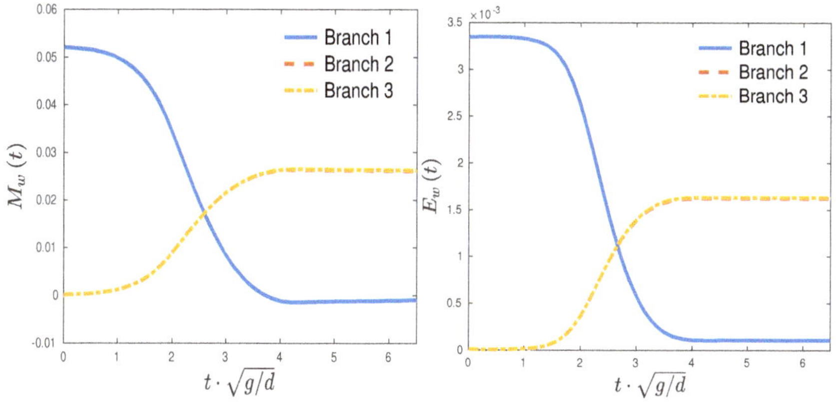

Figure 4. Time evolution of the wave mass M_w (**left**) and the wave energy E_w (**right**) for a wave incident in branch 1 for $a/d = 0.1$.

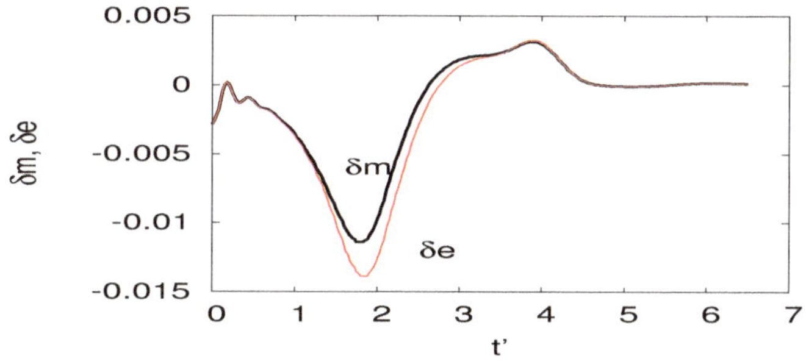

Figure 5. Time evolution of the mass and energy quantities δm (black online) and δe (red online) for $a/d = 0.1$.

We can use them to obtain u_2, h_2. Assume symmetry $h_2 = h_3, u_2 = -u_3$. The balance laws reduce to

$$-h_1 v_1 - 2h_2 u_2 = 0, \tag{45}$$

$$-v_1(gh_1^2 + h_1 v_1^2/2) - 2u_2(gh_2^2 + h_2 u_2^2/2) = 0. \tag{46}$$

Since $v_1^2, u_2^2 \ll gh^2$ we can neglect the terms v_1^2, u_2^2 of the second equation. The resulting relations are satisfied by

$$h_2 = h_1,\ u_2 = -v_1/2, \tag{47}$$

which are the Stoker conditions. These are in good agreement with the simulations as shown by Figure 6.

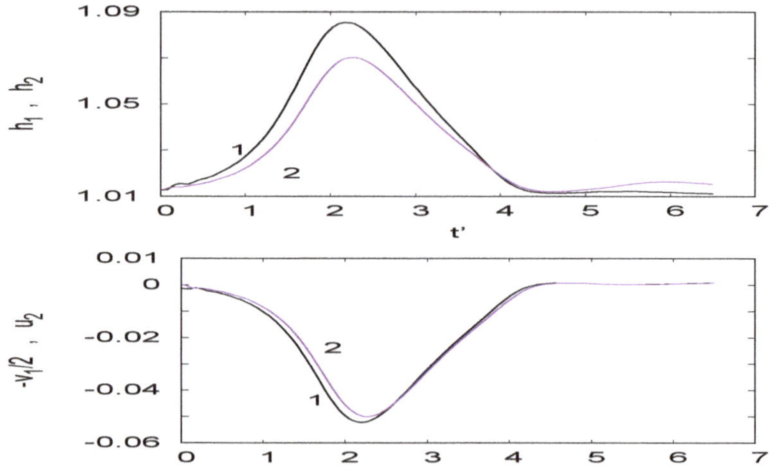

Figure 6. Time evolution of h_1, h_2 (**top**) and $v_1/2, u_2$ (**bottom**) for $a/d = 0.1$.

5.1.2. Very Large Amplitude Waves $a/d = 2$

In this case, 2D effects start to appear. Figure 7 shows a snapshot of the surface elevation h for a wave such that $a/d = 2$. Notice the lump $h \approx 2$ on the edge of the domain.

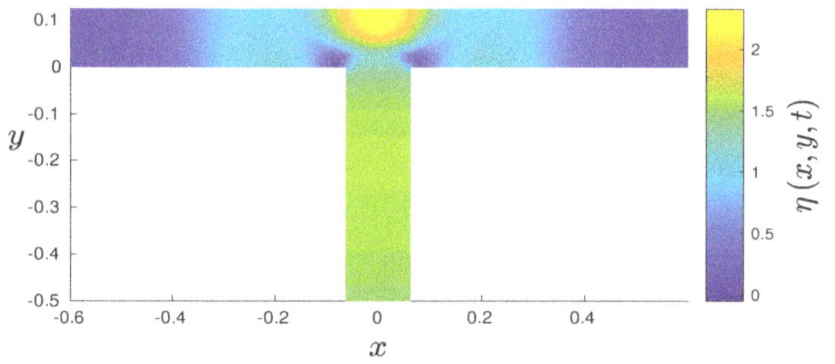

Figure 7. Snapshot of the surface elevation h at time $t = 0.9$ for a wave incident in branch 1 for $a/d = 2$.

Figure 8 shows the time evolution of the wave mass and energy. Despite the evidence of 2D effects, the overall transfer of wave mass and wave energy from branch 1 to branches 2 and 3 does not vary significantly as a/d changes from 0.1 to 2.

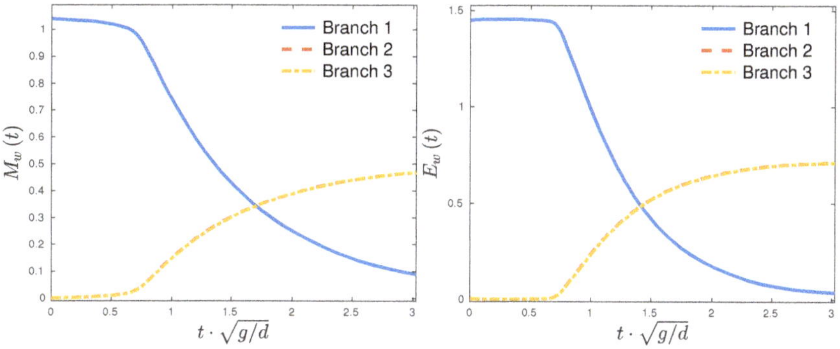

Figure 8. Time evolution of the wave mass M_w (**left**) and the wave energy E_w (**right**) for a wave incident in branch 1 for $a/d = 2$.

Figure 9 shows the time evolution of δm and δe. Notice that the mass relation is better satisfied than the energy relation.

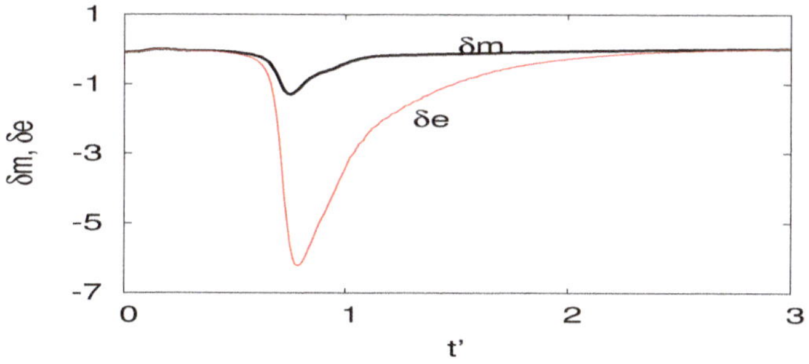

Figure 9. Time evolution of the mass and energy quantities δm, δe for $a/d = 2$.

Again the Stoker relations (47) give a good approximation as shown by Figure 10 which show that $h_2 \approx h_1$ and $u_2 \approx v_1/2$. The price to pay to approximate the 2D situation by a 1D effective model is an energy loss at the junction.

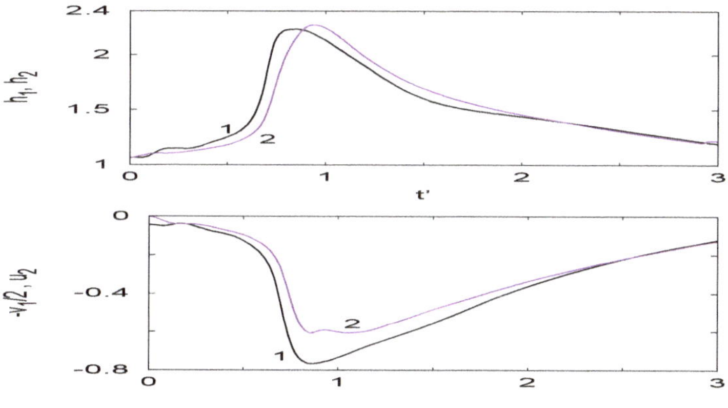

Figure 10. Time evolution of h_1, h_2 (**top**) and $v_1/2, u_2$ (**bottom**) for $a/d = 2$.

Also remark that for the approximation to hold it is crucial that the wave be wider than w and not too fast. If these conditions are not met, h_2 and u_2 will be delayed from h_1, v_1 and will need to describe what happens in the fork. We observed this for a larger channel $w = 1$ and the same parameters.

5.2. Wave Incident into Branch 3

For this configuration, we observe a significant difference in behavior as the wave amplitude increases. Figure 11 shows the time evolution of the wave mass and wave energy for $a/d = 0.1$ (top panels) and $a/d = 2$ (bottom panels). Small amplitude waves get transmitted to branch 1 as much as to branch 2. On the other hand, large amplitude waves are predominantly transmitted to branch 2. The mass entering branch 2 is three times larger than the one entering branch 1; for energies, the factor is six.

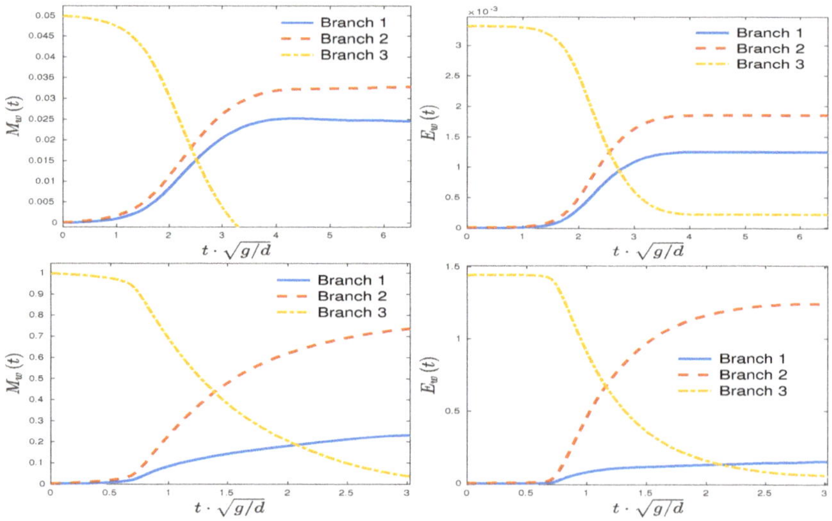

Figure 11. Time evolution of the wave mass M_w (**left**) and the wave energy E_w (**right**) for a wave incident in branch 3 for $a/d = 0.1$ (**top panels**) and $a/d = 2$ (**bottom panels**). Notice the different scales.

5.2.1. Small Amplitude Waves $a/d = 0.1$

First observe that u_1 is non zero and close to v_1. Nevertheless, the mass and energy residuals δm and δe are small as seen in Figure 12. The wave elevation h does not vary much from one branch to the other as seen in the top panel of Figure 13. The velocities u_2 and v_1 verify $u_2 \approx u_3/2, \;\; v_1 \approx u_3/2$. These two results show that the Stoker conditions hold for this small amplitude.

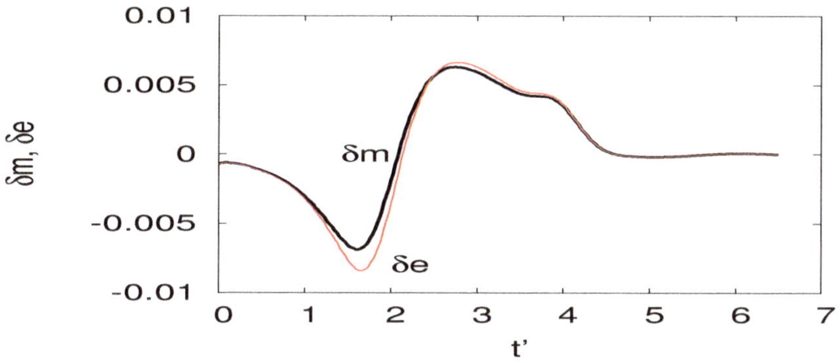

Figure 12. Time evolution of the mass and energy quantities δm (black online) and δe (red online) for $a/d = 0.1$.

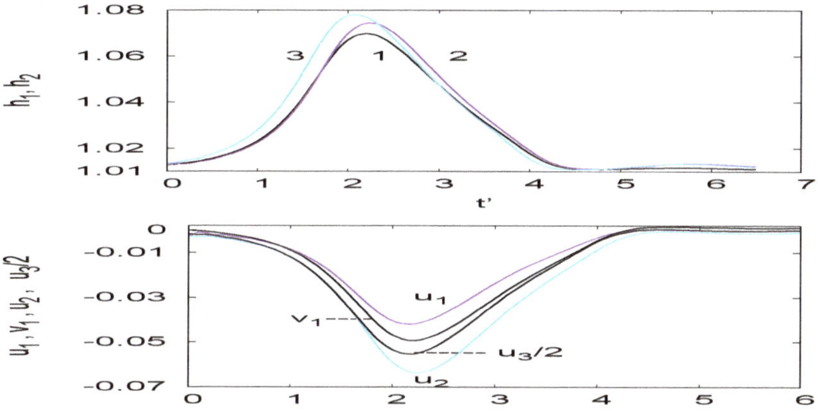

Figure 13. Time evolution of h_1, h_2 and h_3 (top) and u_3, u_2, u_1, v_1 (bottom) for $a/d = 0.1$.

5.2.2. Large Amplitude Waves $a/d = 1$

Figure 14 shows $h(t = 0.8)$ for a wave incident in branch 3 for $a/d = 2$. Notice the complex structure of the flow at the junction. There is some recirculation so that the flow is essentially 2D and not amenable to a 1D reduction.

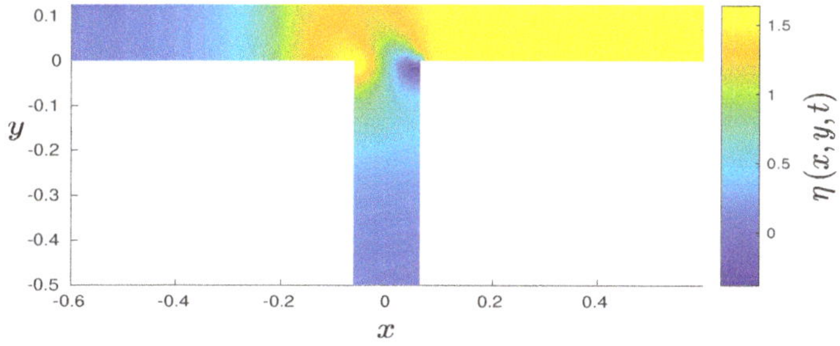

Figure 14. Snapshot of the surface elevation h at time $t = 0.8$ for a wave incident in branch 3 for $a/d = 2$.

Nevertheless, for a smaller amplitude $a/d = 1$, the balance laws (41)–(44) give some insight into the flow. Figure 15 shows the mass δm and energy δe. The mass is much better conserved than the energy.

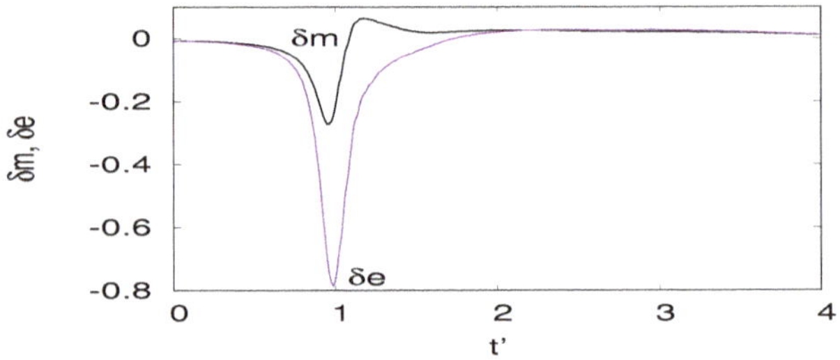

Figure 15. Time evolution of the mass and energy quantities δm (black online) and δe (purple online) for $a/d = 2$.

The momenta (42) and (43) are plotted in Figure 16.

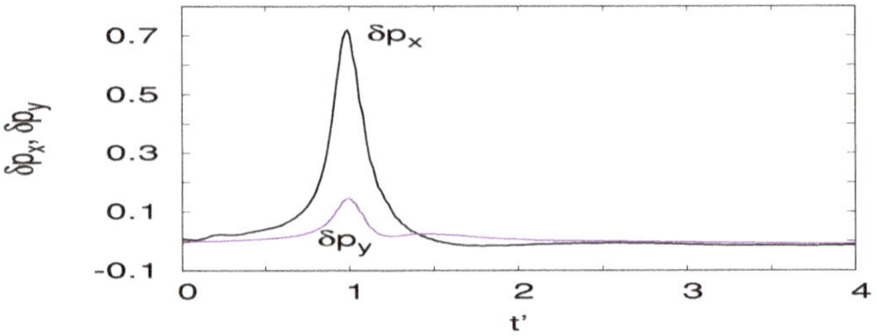

Figure 16. Time evolution of the x and y momenta quantities δp_x (black online) and δp_y (purple online) for $a/d = 1$.

When the wave is coming from branch 3, an obvious solution is

$$v_1 = 0, \quad u_2 = u_3, \quad h_2 = h_3, \quad h_1 = h_2. \tag{48}$$

This is simplistic, in reality $v_1 \neq 0$ but remains small. The horizontal component u_1 is non zero and close to u_2 as shown in Figure 17.

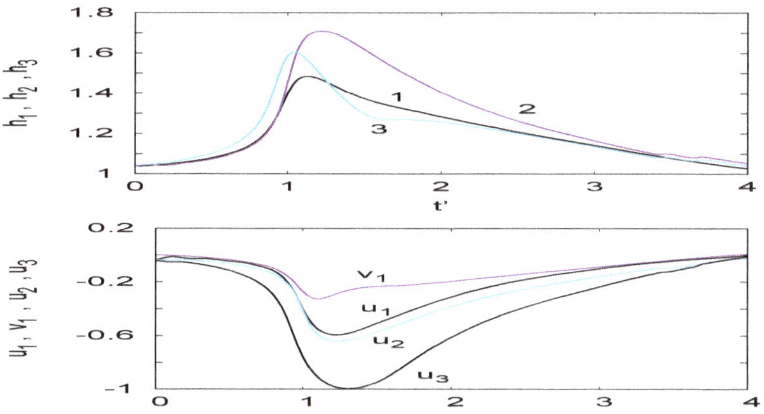

Figure 17. Time evolution of h_1, h_2, h_3 (**top**) and u_3, u_2, u_1, v_1 (**bottom**) for $a/d = 1$.

The mass equation and y momentum equations allow to extract relations between $v_1, h_1, h_2, u_2, v_1, h_3, u_3$. Assuming v_1, u_2, u_3 smaller than h_1^2, h_2^2, h_3^2, we have

$$v_1 = \frac{h_3 u_3 - h_2 u_2}{h_1}, \tag{49}$$

$$h_1 = h_{23}. \tag{50}$$

The quantity (34) in the y component of the momentum is computed from the numerical solution. It is plotted as a function of time together with the estimate

$$h_{23}^i = \sqrt{\frac{1}{2}(h_2^2 + h_3^2)}, \tag{51}$$

in the left panel of Figure 18. As can be seen, the agreement is very good.

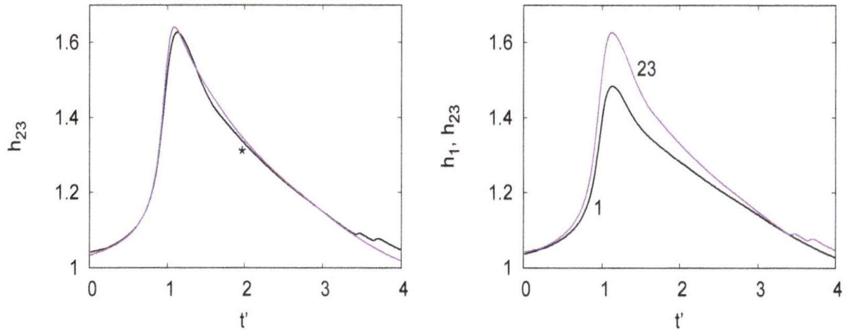

Figure 18. (**Left**) panel, time evolution of the quantity h_{23} (purple online) from (34) obtained from the 2D numerics together with the approximation (51) (black online) indicated by the $*$ symbol. (**Right**) panel, time evolution of h_{23} and h_1.

The velocity v_{1m} given by the mass conservation relation agrees semi-quantitatively with the value v_1 estimated from the 2D numerical solution. Both quantities are plotted as a function of time in Figure 19. Note the delay due to the time the wave needs to propagate from one interface to the other. The y momentum conservation law is not satisfied so that there is no additional equation to estimate u_2.

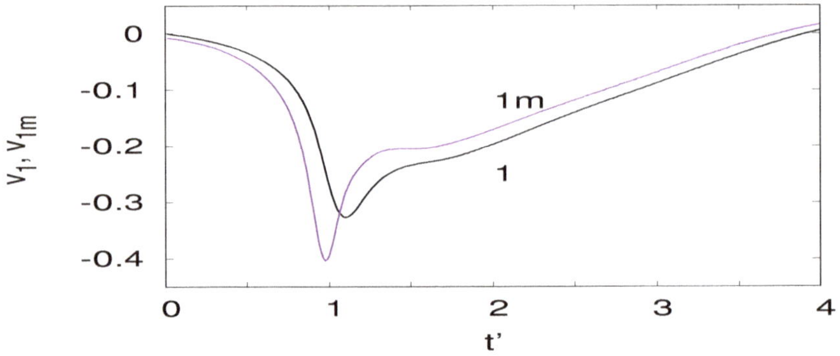

Figure 19. Time evolution of the quantity v_{1m} obtained from the mass conservation law (49) (purple online) and v_1 from the 2D numerical solution (black online).

6. Discussion and Conclusions

The results of the previous section show that for large amplitudes and an asymmetric fork Stoker's interface conditions do not hold and the angle of the fork plays a role. This seems to contradict the findings of Shi et al. [9]. Two reasons show that there is no contradiction. First, the amplitude of our waves ($a/d \approx 1$) are much larger than the ones presented in [9] ($a/d \approx 0.3$) so that nonlinear effects are much stronger in our study. The other point is that the sech^2 initial condition is an exact solution of the Boussinesq equations, but not of the nonlinear shallow water equations. For the Boussinesq equations, we also expect an angle dependence, even for narrow channels, when the amplitude becomes large. To see this, we examine the reduction of the equations for a fork.

The Boussinesq equations read

$$h_t + \nabla \cdot [(1+h)\nabla\varphi] = 0, \tag{52}$$

$$\varphi_t + \frac{1}{2}(\nabla\varphi)^2 + h - \frac{1}{3}(\Delta\varphi)_t = 0, \tag{53}$$

where $h(x,y,t)$ is the water elevation. The velocity potential $\varphi(x,y,t)$ is such that $(u,v)^T = \nabla\varphi$. The boundary conditions are non slip $\nabla\varphi \cdot \mathbf{n} = 0$. Integrating the equations on the fork domain $\mathcal{F}$ (left panel of Figure 1) we get

$$\partial_t \int_{\mathcal{F}} h dxdy - \int_{IA \cup CD \cup FG} (1+h)\nabla\varphi \cdot \mathbf{n} ds, \tag{54}$$

$$\partial_t \int_{\mathcal{F}} (\varphi - \frac{1}{3}\Delta\varphi) dxdy + \int_{\mathcal{F}} (\frac{1}{2}(\nabla\varphi)^2 + h) dxdy = 0. \tag{55}$$

Neglecting the time evolution in the fork region, we get the following interface conditions

$$(1+h_1)u_1^{\|} + (1+h_2)u_2^{\|} + (1+h_3)u_3^{\|} = 0, \tag{56}$$

$$\int_{\mathcal{F}} \frac{1}{2}\left[(\nabla\varphi)^2 + h\right] dxdy = 0. \tag{57}$$

Note how the first equation reduces to Kirchhoff's law for small h. The second equation contains is an integral over the whole domain and depends on the angle of the fork. For small angles, we can assume that $\nabla\varphi = u^{\|}$ so that the conditions reduce to

$$(1+h_1)u_1^{\|} + (1+h_2)u_2^{\|} + (1+h_3)u_3^{\|} = 0, \tag{58}$$

$$\frac{1}{2}(u_1^{\|})^2 + h_1 + \frac{1}{2}(u_2^{\|})^2 + h_2 + \frac{1}{2}(u_3^{\|})^2 + h_3 = 0. \tag{59}$$

Not surprisingly, these conditions are very close to the ones obtained by Nachbin and Simoes [10], except for the Jacobian of the conformal transformation.

To conclude, we studied the propagation of shallow water waves in a fork between three narrow channels. We considered both the 2D numerical solution and a homothetic reduction procedure that gives coupling conditions at the interface. For such narrow widths, the delay experienced by the wave is negligible so that one can envision describing the junction by an effective 1D PDE model.

Our reduction enabled us to derive balance laws for the mass, momenta and energy of the flow across a general junction. For small amplitude waves, these laws reduce to the commonly used Stoker jump conditions, giving these a formal justification. We verified these Stoker conditions on the 2D numerical solutions of the shallow water equations for symmetric and non symmetric conditions. Then, the angle of the junction does not play any role. This happens also for a general nonlinear wave equation; we had seen this a previous study for the particular case of the sine-Gordon equation [2].

For large amplitude shallow water waves, the situation depends on the symmetry of the fork. For a symmetric fork, the Stoker conditions are approximately verified. This is explained by the strong constraint imposed by the symmetry. Then, the only solution of the balance laws corresponds to the Stoker conditions. When the fork is non symmetric as in our case 2, more information is needed about what happens inside the fork. The quantities $u_i^{\|}$, $i = 1,2,3$ are velocities projected in the direction of the branches and this projection leads to a loss of information. Far from the junction, the flow is quasi-1D so that not much is lost. On the contrary, inside the junction, the flow is full 2D. A possible solution, to be studied in the future would be to use the full conservation law including the time dependent term. Then we would introduce a fictitious node inside the junction and couple it to the boundaries using average differential equations obtained by integrating (9)–(11) on the fork domain.

Author Contributions: J.-G.C. calculated the conservation laws, analyzed the numerical data and wrote the manuscript. D.D. performed the 2D simulations of wave propagation across the fork with VOLNA code. He also participated in the analysis of the numerical results. B.G. studied the stucture of the conservation laws and the numerical results.

Funding: This research was funded by ANR grant "Fractal grid" and by the CNRS through the project PEPS InPhyNiTi "FARA".

Acknowledgments: The authors thank Tim Minzoni and Nick Ercolani for useful discussions. We also thank the Centre Régional Informatique et d'Applications Numériques de Normandie for the use of its computers.

Conflicts of Interest: The authors declare no conflict of interest.

References

1. Bressan, A.; Čanić, S.; Garavello, M.; Herty, M.; Piccoli, B. Flows on networks: Recent results and perspectives. *EMS Surv. Math. Sci.* **2014**, *1*, 47–111. [CrossRef]
2. Caputo, J.-G.; Dutykh, D. Nonlinear waves in networks: model reduction for the sine-Gordon equation. *Phys. Rev. E* **2014**,*90*, 022912. [CrossRef] [PubMed]
3. Hadamard, J. Leçons sur la géométrie élémentaire. *Librairie Armand Colin* **1906**, *17*, 103–209.
4. Stoke, J.J. *Water Waves: The Mathematical Theory with Applications*; Wiley-Interscience: Hoboken, NJ, USA, 1992.
5. Jacovkis, P.M. One-dimensional hydrodynamic flow in complex networks and some generalizations. *Siam J. Appl. Math.* **1991**, *51*, 948–966. [CrossRef]
6. Holden, H.; Risebro, N.H. Riemann problems with a kink. *Siam J. Appl. Math.* **1999**, *30*, 497–515. [CrossRef]
7. Landau, L.; Lifchitz, E. *Cours de Physique Théorique: Hydrodynamique*, Ellipses: Moscow, Russia, 1990.

8. Georg Schmidt, E.J.P. On Junctions in a Network of Canals. In *Control Theory of Partial Differential Equations*; Chapman and Hall/CRC: Boca Raton, FL, USA, 2005; pp. 207–212.
9. Shi, A.; Teng, M.H.; Sou, I.M. Propagation of long water waves through branching channels. *J. Eng. Mech.* **2005**, *131*, 859. [CrossRef]
10. Nachbin, A.; Simoes, V.S. Solitary waves in forked channel regions. *J. Fluid Mech.* **2015**, *777*, 544–568. [CrossRef]
11. Exner, P.; Post, O. Quantum networks modelled by graphs. *AIP Conf. Proc.* **2008**, *998*, 1.
12. Post, O. Spectral Analysis on Graph-Like Spaces. In *Lecture Notes in Mathematics*; Springer: Berlin, Germany, 2012.
13. Dutykh, D.; Caputo, J.-G. Wave dynamics on networks: Method and application to the sine-Gordon equation. *Appl. Numer. Math.* **2018**, *131*, 54–71. [CrossRef]
14. Dutykh, D.; Katsaounis, T.; Mitsotakis, D. Finite volume schemes for dispersive wave propagation and runup. *J. Comput. Phys.* **2011**, *230*, 3035–3061. [CrossRef]
15. Dutykh, D.; Poncet, R.; Dias, F. The VOLNA code for the numerical modeling of tsunami waves: Generation, propagation and inundation. *Eur. J. Mech. B* **2011**, *30*, 598–615. [CrossRef]
16. Shampine, L.F.; Reichelt, M.W. The MATLAB ODE Suite. *SIAM J. Sci. Comput.* **1997**, *18*, 1–22. [CrossRef]
17. Serre, F. Contribution à l'étude des écoulements permanents et variables dans les canaux. *La Houille Blanche* **1953**, *8*, 374–388; 830–872. [CrossRef]
18. Dutykh, D.; Clamond, D.; Milewski, P.; Mitsotakis, D. Finite volume and pseudo-spectral schemes for the fully nonlinear 1D Serre equations. *Eur. J. Appl. Math.* **2013**, *24*, 761–787. [CrossRef]

MDPI
St. Alban-Anlage 66
4052 Basel
Switzerland
Tel. +41 61 683 77 34
Fax +41 61 302 89 18
www.mdpi.com

Symmetry Editorial Office
E-mail: symmetry@mdpi.com
www.mdpi.com/journal/symmetry

www.ingramcontent.com/pod-product-compliance
Lightning Source LLC
Chambersburg PA
CBHW051909210326
41597CB00033B/6079